Photo by William Pullman

Michael Wheeler is an associate professor of law at New England School of Law, in Boston. The author of *No-Fault Divorce,* he has also written for *The Atlantic Monthly, New Times,* and a number of legal journals. He and his wife, Candace, live in Cambridge, Massachusetts.

Lies, Damn Lies, and Statistics

By the same author

No-Fault Divorce

Lies, Damn Lies, and Statistics

The Manipulation of Public Opinion in America

Michael Wheeler

LIVERIGHT

NEW YORK

Copyright © 1976 by Michael Wheeler
FIRST EDITION

ALL RIGHTS RESERVED
Published simultaneously in Canada
by George J. McLeod Limited, Toronto

Library of Congress Cataloging in Publication Data

Wheeler, Michael, 1943–
Lies, damn lies, and statistics; See Shp·

Bibliography: p. 288–289.
Includes index.
1. Public opinion polls.   2. Public opinion—
United States.   3. Voting research—United States.
I. Title.
HM261.W48 1976   301.15′4′0973   76–6886
ISBN 0–87140–621–7

PRINTED IN THE UNITED STATES OF AMERICA

1   2   3   4   5   6   7   8   9

For L. David Otte,
whose enthusiastic support of this book went
far beyond that required of a literary agent
or expected of a friend.

Contents

"There are three kinds of lies—lies, damned lies, and statistics."
—attributed to both Benjamin Disraeli and Mark Twain

Acknowledgments

I am much indebted to the scores of pollsters, politicians, and reporters who generously shared their insights and experiences with me. Their observations greatly enriched this book. Unfortunately, I cannot acknowlege the contribution of each person whom I interviewed, as there simply is not space here to list all the names.

There are certain other people whose help was so important that they must be thanked individually. Rob Duboff, long a student of polling and now a pollster himself, supported my efforts from the start, by providing much valuable information and by being constantly available for me to test my ideas. Ned Arnold, who signed up the book, was always enthusiastic in his editorial prodding.

Representative Michael Harrington was instrumental in arranging interviews for me with many of his colleagues in Congress. Senator and Mrs. Benjamin Smith II were most helpful on the same score, and they also provided the haven in which I wrote most of the book. In the face of tight deadlines, the staff at New England School of Law, especially Tia Hawkes, helped produce the manuscript quickly and efficiently.

Several friends, Tom Horst, André Mayer, and Lurana Snow, read an earlier draft, and their suggestions helped me focus my thesis and present it more clearly. I am particularly grateful to my wife, Candace, and my parents, Harry and Erma Wheeler, who not only read the book as it evolved through various stages, but who put up with my odd hours and frazzled nerves.

Finally, I must thank several people whose identities I promised to protect: a pollster who provided me with important leads, and poll-takers who, contrary to their employers' policies,

allowed me to follow them in their rounds. These people un-
selfishly supplied me with some of the most useful information
that I obtained. I was encouraged that they shared my enthusiasm for the project and deeply touched that they had confidence
in me to keep their trust.

Michael Wheeler

Cambridge, Massachusetts

Preface

At any moment, day or night, Gallup poll-takers are knocking on doors, gathering public opinion. Even when it is after midnight in the United States, interviewers for Gallup's affiliated firms are at work in countries such as Greece, Iran, India, and Japan. Around the clock, Gallup is engaged in a global pulse-taking.

Pollster Louis Harris has established his Harris Alert System so that he can activate his field operation almost instantly and thus get up-to-the-minute readings on public reaction to late-breaking events, whether they involve presidential politics or international conflict.

Three hundred sixty-five nights a year in Swarthmore, Pennsylvania, specially trained operators for the Sindlinger Company dial randomly selected telephone numbers throughout the country to collect people's opinions about the economy. Private companies pay up to five thousand dollars to see the results of the surveys. Over the last twenty-five years Sindlinger has interviewed four and a half million people.

The A. C. Nielsen Company, which determines the television ratings, has so perfected its survey techniques that it has completely dispensed with interviewers. Television sets in the Nielsen sample are wired to a central computer which silently monitors the viewing habits of each selected household.

For every well-known pollster like Gallup, Harris, Sindlinger, or Nielsen, there are literally hundreds of others engaged in the same sort of activity. Some do published newspaper polls, while others specialize in private surveys for politicians or businessmen. There are more than a thousand polling organizations in the United States, and their total revenue is estimated to be at least half a billion dollars a year, perhaps much more.

The breadth and intensity of modern public opinion polling is matched by its influence. Pollsters have become the *vox populi*, and their work touches all our lives.

The most conspicuous influence of polls is on politics. Almost every candidate for public office, national or local, takes a poll before jumping into the race. Ronald Reagan retained the aptly named polling firm Decision Making Information to take a national survey to assay his chances before he decided to challenge President Gerald Ford for the Republican nomination. Early public opinion polls certify a chosen few as "serious candidates," and as such, these people are blessed with campaign contributions, endorsements, and substantial media coverage. Those who are labeled "also rans," often long before the campaigning actually begins, face an upstream swim.

The pollsters are defensive about the charge that they are tampering with the electoral process, and almost unanimously deny that their polls create bandwagons in favor of those who are given early leads. Politicians, however, disagree. Senator Hubert Humphrey says: "We have to face the fact that polls do have a political impact. Politicians know it. Why else would they spend money for a poll and then leak it to the press, if it didn't help?"

Public opinion polls are also becoming increasingly important in government policy-making. Careful politicians have always put a finger in the wind before taking positions on controversial issues; many consult the newspaper surveys of Gallup and Harris, while others sponsor private polls of their constituents. In recent years, however, this dependence on pollsters has become more overt; public-opinion polling is becoming an institutionalized part of the democratic process.

The administrative branch of government has for many years used polls to fashion policy; the legislative branch is following suit. In September 1975, for example, eight pollsters were called to testify before the Senate Foreign Relations Committee about American attitudes on foreign policy. The witnesses were a veritable all-star team of pollsters; among them were Louis Harris, Burns Roper, Daniel Yankelovich, and George Gallup, Jr., standing in for his father.

One month later the Joint Economic Committee of Congress heard six pollsters testify on American opinion about business, government, and the economy. Harris again was there and Gallup was represented by the vice-president of his organization. The appearance of the major pollsters on Capitol Hill is becoming commonplace.

Should we celebrate this development? (Gallup reports that 67 percent of the population wish leaders would heed the polls.) If Congress is really listening to the people, then we should be gratified. Instead, however, only a handful of men, the pollsters, are chosen to speak for us, and they do it poorly.

Polls on issues are often misinterpreted and some are deliberately rigged, yet they are all treated with unquestioning respect. For years misleading polls which seemed to indicate popular support for the war in Vietnam stifled public criticism of United States policy. Contrary to the conventional wisdom, public opinion polls did not hasten the resignation of Richard Nixon; rather, they delayed it. Had it not been for other events, the polls might have kept him in office.

The influence of polls extends far beyond elections and government policy-making. The little boxes in twelve hundred homes selected by the Nielsen Company register an electronic thumbs up or down on the television programs which will be watched by the entire country. A shift of a rating point or two can determine whether a program lives or dies, yet each rating point represents only a dozen Nielsen households. If just twelve families, scattered across the United States, happen to switch off a particular program, it may be cancelled even though tens of millions of people watch it. Because advertising revenue is directly tied to ratings, a network which slips a single point over the course of a year stands to lose as much as twenty million dollars!

Hundreds of millions of dollars are spent each year on market research, less to develop new and better products than to fashion more reassuring corporate images. The use of polls is spreading into other areas, such as law. In one obscenity case, a lawyer commissioned a sixty-five-hundred-dollar survey to determine community attitudes about sexually explicit material. Other lawyers have used polls to see whether or not their clients should seek a jury trial.

At the same time that public opinion polls are becoming increasingly influential in politics, business, and elsewhere, there are disturbing signs that the surveys themselves may be fundamentally flawed.

In late 1975 the *New York Times* reported that opinion researchers are encountering great resistance from the public. More and more people are refusing to be interviewed by the pollsters. Mervin Field, the head of the California Poll, was quoted as complaining: "Twenty years ago we could figure on getting 85 percent with reasonable effort. Now we're hard pressed to get 60 percent."

To make matters worse, changing work and leisure habits make it hard to find people at home. Even the academic researchers, who sometimes take the trouble of going back to a house seven or eight times to try to interview an occupant, are finding that in some localities they are lucky if they can complete half their assignments.

The high rate of refusals and not-at-homes is just one of the many technical problems which pollsters face, but it strikes at the very heart of public opinion polling, for if a great number of people are not answering the pollster's knock on the door, then opinion surveys cannot be truly representative of the entire population.

People who refuse to talk to pollsters are likely to be different from those who do. Some may be fearful of strangers and others jealous of their privacy, but their refusal to talk demonstrates that their view of the world around them is markedly different from that of those people who will let poll-takers into their homes. Similarly, people who have to hold down two jobs have quite different experiences and attitudes from those of people who are habitual homebodies.

The pollsters themselves concede the magnitude of the problem, but have yet to come up with a solution. Irving Crespi, vice-president of the Gallup Organization, has stated: "I wonder whether we are reaching the point where we are saturating the public. It is a very serious problem that threatens the long-term viability of the survey profession."

The viability of public opinion polling may well be threatened by this and other problems, but you would not know it from reading the published polls of Gallup, Harris, and the rest. The reader of opinion polls is given no hint that they may not be trustworthy. The press in particular has been oblivious to the pitfalls of polling.

On the very day that the *New York Times* reported the considerable difficulties that pollsters are now encountering in trying to interview people, it also published on the front page a survey on public opinion about the economy. The accompanying story took up more than fifty column inches, but there was no indication of what portion of the sample could not be located or how many people refused to talk to the pollsters. Three times in the next two weeks the *Times* ran public opinion polls on the front page. Not once was there any acknowledgment that methodological problems necessarily make the results open to question! If the *Times* does not heed its own warnings about the viability of polls, it should not be surprising that other papers are equally uncritical about what they publish.

The question of the trustworthiness of the polls goes beyond questions of methodology or careless coverage by the press. So much now depends on the opinion polls—political candidacies, policies, profits—that there is great pressure on the pollsters to manipulate their surveys. It is not hard to rig a poll. According to Washington, D.C., pollster Peter Hart, simply by subtly altering the wording of a question "you can come up with any result you want."

Hart is a person of high integrity, but not all of his competitors are as principled. Many pollsters will come up with a poll which says whatever the client wants, so long, of course, as the client is willing to pay for it. Taking note of this sort of activity, Burns Roper has concluded, "While we are probably the newest profession in existence, we have managed in a few short years to take on many of the characteristics of the world's oldest profession."

This book is about public opinion polls—how they are used and abused. The polls have come to have a far-reaching and often dangerous impact, largely because the pollsters have intimidated most of their potential critics. The pollsters would have us believe that to criticize their calling is to oppose democracy, but, their claims to the contrary, public opinion and public opinion polls are two quite different things. The pollsters have also used pseudo-scientific jargon to ward off criticism. Most of us are impressed by the definiteness and clarity of statistics but feel utterly incompetent to evaluate them.

It is necessary to demystify and demythologize the polling process, to demonstrate in non-technical terms just what polls can and cannot do. Even perfectly constructed surveys have inherent limitations, and pollsters often have to compromise the accuracy of a poll to cut costs. Moreover, not all pollsters are equally skilled; some make serious mistakes.

It is also necessary to know the pollsters, these invisible men who wield so much power. How do they profit from polling? What are their prejudices? It would be wrong, however, to regard the pollsters as villains. If, as Roper suggests, they can be prostitutes, then somebody must be buying their services. How have we let such enormous power fall into so few hands? Why are we so bedazzled by names and numbers?

It is not enough to raise the dangers which polls pose for our political and social system. Greater awareness of the harm that can be caused by inaccurate or rigged polls may be of some value, but more must be done to bring the pollsters into line. What remains to be seen is whether they have so firmly insinuated themselves into our national life that their pretensions cannot be deflated and their power defused.

Lies, Damn Lies,
and Statistics

1

A Prologue
Nixon and the Pollsters

In late 1970 Richard Nixon instructed Charles Colson to meet with pollster Louis Harris to find out why Harris' ratings of the president were consistently more negative than were George Gallup's. Colson's assignment was the first step in what became an unprecedented attempt by the White House to manipulate the public opinion polls. The operation is a revealing postscript to the Nixon administration, but, more important, it shows how willingly the supposedly impartial arbiters of public opinion play behind-the-scenes politics.

Nixon is now in exile in San Clemente, Colson has served a prison sentence, but the pollsters—who chart the ups and downs of the public pulse, and prognosticate on coming elections—are still very much with us.

Nixon had reason to worry about Lou Harris' polls. During 1970 Harris had the president's support slipping from a high of 58 percent down to 48 percent by the time Colson was sent to talk with him. By contrast, Gallup had approval of the president in a much higher range, 68 percent to 53 percent. During the year Harris' rating had run anywhere from one to nine points below Gallup's.

The difference, Nixon suspected, was not a statistical quirk but

the result of a deliberate vendetta on Harris' part. Harris had the
reputation of being a Democratic partisan, having made his mark
by polling for John Kennedy in 1960. In *Six Crises* Nixon bitterly
recalls how Harris' polls were used to fuel a propaganda "blitz"
against him in the late stages of that campaign. Two years later
Nixon ran for the governorship of California and again encountered Harris, who this time was polling for Pat Brown. Nixon
was convinced that Harris was leaking erroneous polls to hurt
him.

By 1968, when Nixon next was a candidate, Harris had given
up private political work in favor of his syndicated newspaper
survey, yet the pollster still managed to step on Nixon's toes.
Throughout the campaign, Harris gave Nixon a significantly
smaller lead over Hubert Humphrey than did Gallup, and in a
controversial election-eve poll Harris projected a 45 to 41 percent
Humphrey margin over Nixon (with the rest going to George
Wallace). Nixon won the election, of course, but that did not
make his supporters feel the least bit charitable toward Harris.
Herb Klein, then Nixon's director of communications, told a
meeting of newspaper editors, "Harris ought to be put out of
business."

Thus, in late 1970 the stage was hardly set for a congenial
meeting between Colson and Harris. Colson, however, held an
important trump card. Harris' polling organization had recently
been purchased by the brokerage firm Donaldson, Lufkin & Jenrette. Colson, as it happened, was a friend of one of the partners,
Dan Lufkin. Colson recalls, "Harris was not very friendly in
those days, but Lufkin owned him, and that enabled me to get an
audience where at least he would talk to me civilly."

Colson came to New York armed with reams of statistics
which ostensibly proved that the Harris Survey had a consistent
anti-Nixon tilt. Chief among his arguments was the contention
that Harris' way of testing presidential support was overly negative. Gallup simply asked, "Do you approve or disapprove of the
way Nixon is handling his job?" By contrast, Harris asked, "How
would you rate the job President Nixon is doing as president—
excellent, pretty good, only fair, or poor?" Those people who said
"excellent" or "pretty good" were tabulated as giving the president a positive rating, while those who selected the other two

categories were counted as being negative.

Colson says: "That's one of the ways I thought he slanted it. If a Vermonter says 'fair,' that's quite a tribute. For a lot of people, 'fair' at least is neutral, not negative." Colson states that he was able to convince Lufkin, who sat in on the discussion, that Harris' poll was biased, but that Harris himself did not budge. In his book *The Anguish of Change*, Harris alludes briefly to the meeting and describes the exchange as "pointless." During the next six months, Harris' ratings of Nixon continued to run from five to eight points lower than Gallup's.

Colson was both right and wrong in his criticism. There is no authoritative way to test popular support for a president. Indeed, the results turned up by any single question, be it Gallup's or Harris', are bound to be crude. Colson was correct in seeing that Harris' method of tabulation would be expected to produce a lower measure than did Gallup's, but the two ratings moved up and down more or less in unison. One could not be said to be more accurate than the other.

Although Colson failed to convert Harris that day, his visit marked an important turning point because it opened up communications between the White House and the pollster. In the course of subsequent phone calls and meetings, the relationship became warmer. In Harris' words, "a truce of sorts" was worked out in which he agreed to provide the White House with pre-publication reports of his surveys; in addition, he made himself available to discuss any of his polls. Harris had provided the same service for both the Kennedy and Johnson administrations. In return, according to the pollster, "the White House agreed not to send out any emissaries to 'get Harris.' "

Perhaps it should not be surprising that Harris, a former new frontiersman, would cooperate with the Nixon administration, for he has always been irresistibly drawn toward power. He takes pride in the fact that in 1960 he sat on "the inner strategy committee" with Joseph, John, and Robert Kennedy (a claim some former Kennedy aides dispute). He says that one of his motivations in publishing his newspaper column is to "have some impact with the movers and shakers of the world."

The White House, through Colson, exploited the Harris relationship, calling on him often. Colson says: "We entered sort of an era of cooperation with Lou. We got an awful lot of very helpful advice." Colson was delighted to have the inside word on public opinion, and Harris was pleased to have some influence. Few people in politics or polling were aware that John Kennedy's former pollster was becoming increasingly cozy with Nixon.

In the spring of 1971 John Ehrlichman wanted his Domestic Council to commission an intensive poll on American attitudes about a host of social issues. Government agencies commonly hire pollsters to conduct surveys to be used in planning and making policy, but a poll for the White House itself, particularly with an election year approaching, obviously would be useful for political strategy as well. For a time, the results of the poll were classified as "secret."

A contract for a government poll is a plum which is not handed out casually, and if there was ever an administration which rewarded its friends and got back at its enemies, it was Nixon's. According to Colson: "Ehrlichman's plan for the poll came up in a conversation with the president. I made the point that Harris had been giving us a lot of helpful information. This might be a prestige poll for him. Let's let him do it."

Long before that time, H. R. Haldeman had blacklisted Harris to keep him from getting any government work, a fact which horrified Colson: "Despite what you may have read about me, I didn't play the game that way." Ethics aside, Colson was shrewd enough politically to know that you do not make friends with a powerful pollster by cutting him out of profitable business. Nixon ultimately agreed with Colson, and decided that Harris should be given the contract.

The deal was consummated in a telephone conversation which Harris himself regarded as peculiar. According to the pollster, Nixon called and asked whether he would be willing to do the study. "I said yes, but stressed that it had to be objective, that we had to have final say over what went into it." Nixon replied: "That's no problem, but I do have to ask you one question. In my view you're the best poll-taker in the business, but I've always thought you were out to get me." Harris repeatedly assured him that he was not, and Nixon concluded the conversation by saying, "Well, I just want you to know that I am the one who selected you to do this Domestic Council study."

It is startling that Harris accepted the offer with no apparent misgivings. When he went into newspaper polling in 1963, he announced that he was forsaking all work for private political clients. "It has always been my view that major contributions could be made in both the public and private areas of political research, but that the two, in the end, are of necessity mutually exclusive." Harris obviously could not purport to write an objective, unbiased newspaper column while still in the pay of a string of Democrats. What would be accurate news for his column might sometimes be bad news for his private clients. Polling for the White House in 1971, especially with Nixon preparing to run for re-election, was no less compromising.

The question of conflict of interest was not just a matter of appearance. During the spring of 1971, when negotiations for the Domestic Council poll were going on, a remarkable shift occurred in Harris' measurement of presidential support. In March he had reported Nixon's support at 41 percent (compared to Gallup's figure of 50), but in April it jumped to 46 percent, in May to 47 percent, and by June it was up to 50. During those same three months, Gallup actually recorded a drop of two points, so that for the first time Nixon was doing better in Harris' poll than in Gallup's!

What accounts for this sudden turnaround? Neither Harris nor Gallup altered their respective questions in the slightest. Some people in the White House were convinced that Harris altered his figures to make Nixon look better than before. Those in the administration may have their own reasons for wanting to discredit Harris.

Colson prefers to put it a bit more gently. "I happen to like Lou as a person. I don't think he's a good pollster or a bad pollster. He is a pollster who does put a slant into his polls by the way he asks questions. After I developed a good personal relationship with him, I discovered remarkably that we started doing better —Harris' polls started coming more in line with Gallup's. I'm not going to say that Lou shaded anything, but there was a change."

The case against Harris must necessarily be circumstantial. If there was an agreement to rig the poll, it certainly was not put in writing, and conversations about such things are necessarily oblique, witness perhaps Nixon's telephone call to Harris.

Before Harris began dealing with the administration, his ratings of Nixon were invariably lower than Gallup's, sometimes markedly so. Thereafter, the disparity vanished. For all but three weeks of Nixon's last year in office, Harris showed greater support for Nixon than did Gallup.

Other Harris measurements of Nixon's strength showed similar improvement. Unlike four years earlier, Harris' 1972 presidential polls were in accord with Gallup's throughout the campaign; at one point Harris had Nixon leading McGovern by thirty-four points. Perhaps most revealing were Harris' 1973 and 1974 surveys on impeachment, which were very slow to register public support for congressional action. It was not until June 1974—a full year after the Ervin committee hearings and more than six months after the firing of Archibald Cox and the discovery of the eighteen-minute gap in the famous tape—that Harris finally reported that a majority of Americans favored impeachment.

But were his figures accurate—were we really so slow to demand a congressional investigation? Clearly not. Harris' way of asking the impeachment question made us seem more reticent than we actually were. Harris had his interviewers ask people whether they wanted Nixon "impeached and removed from office." Impeachment and removal, of course, are two quite different things. In essence, Harris was asking not whether Nixon should simply be tried, but whether he should be tried and hanged!

Pat Caddell did a lot of polling for Democratic senators and congressmen in 1973 and 1974, and is convinced that Harris' surveys, and others like them, had a stagnating impact on the impeachment process. "There were a lot of people in the corridors on the hill who weren't going to take any action until the majority spoke. The people, in turn, were looking for leadership."

If Harris had asked the right questions, this attitude would have been clearer at a much earlier date. Peter Hart, who polls both for the *Washington Post* and for private candidates, says, "The bottom fell out for Nixon long before any of the newspaper polls said it." In his private surveys, Hart asked, "If Congress had to vote today on whether the Senate should hold a trial to determine President Nixon's innocence or guilt on the charges related to Watergate, would you want you congressman to vote for or against holding this trial?" Hart discovered that when the question was posed that way, far more people expressed support for the impeachment process. Surveys by Caddell and Burns Roper uncovered the same fact.

Indeed, Harris himself inadvertently demonstrated the bias of his own question in a survey he released on the eve of Nixon's resignation. For the first time he reported how people responded to the proposition that "the House of Representatives should vote to impeach President Nixon so that he can be tried in the U.S. Senate." Sixty-six percent said they supported such action, yet when the same people were asked Harris' former question only 56 percent said that Nixon should be "impeached and removed."

Harris is known in the polling profession as one of the most artful draftsmen of questions, yet until the end he persisted in using a question which significantly understated public support for impeachment.

It is small wonder, then, that the White House regarded Harris as a most valuable ally and rewarded him by allowing him to do other government work. Privately, Harris was providing them with valuable political intelligence, and publicly, his newspaper polls were far kinder to Nixon than they had been before. The administration wanted to exploit the relationship still further by using Harris' private surveys for propaganda, and it was here that the pollster balked, for it would have meant that his clandestine association would have become known.

Harris' sensitivity on this score was evidenced during his work for the Domestic Council in 1971. Though he had told Nixon that he must have final control over the questionnaire, there actually was a great deal of give and take between the White House and the pollster over just what questions were to be included.

Just before the final questionnaire was to be printed, the White House sent down some additional questions which they insisted be asked. The agree/disagree statements were clearly designed to elicit pro-administration responses: "Too many people are constantly trying to run this country down; instead, they should be positive and build up the country." "The government should get rid of its giveaway programs and reduce taxes."

People on Harris' staff were convinced that the questions had been written by Nixon himself; Colson denies that this was the case, but other White House sources say that it might have been true. In any event, the White House's motivation was clear: they wanted to be able to publicize a Harris Survey showing great support for administration policies. Because of Harris' old Democratic ties, his work would have been given much more credence than a survey by a sympathetic Republican pollster.

Harris cleverly protected himself by inserting some agree/disagree questions of his own: "Honest criticism of the country can often lead to real progress." "The day the U.S. doesn't have compassion for the poor and unfortunate is the day that we've lost our national character." For every White House question which would produce a supposed mandate to crack down on dissenters, Harris had another which would show support for their rights.

In fairness, it must be noted that Harris denies that this episode ever happened, but his denial is consistently contradicted by the accounts of others who were involved, including not just people in the White House but those on Harris' own staff. The actual questionnaire does indeed include the questions quoted above, as well as others of the same ilk.

The Harris connection was only one strand in a growing web of associations that the Nixon White House developed with pollsters. Colson says, "I always had an interest in polls, and found them to be extremely important in politics, both ways, reading public opinion and having an impact in their own right." He reached out to other pollsters, and hardly anyone of consequence escaped his attention. Some were more receptive than others. Daniel Yankelovich only spoke to him by phone, but Samuel Lubell visited the White House for a consultation. Colson became the de facto minister of polls. His vast store of survey information and his influence with the pollsters gave him direct access to Nixon, and helped move him into the inner circle of presidential power.

Though Colson swears that his reputation was undeserved, he has commonly been depicted as the White House heavy.

Theodore H. White, for example, has written, "To gain his ends, Colson had absolutely no scruple—leaks, plants, forgeries, lies were all part of the game as he played it." According to columnist Mary McGrory, Colson rigged a special newspaper poll on the Vietnam war by buying up thousands of papers and flooding the editors with questionnaires filled out in favor of Nixon's policies.

Yet in spite of his unscrupulous reputation, Colson is remembered by the pollsters he contacted as a most genial man. Irwin Harrison, who polls for the *Boston Globe*, remembers that Colson would track him down by telephone no matter where he happened to be in order to get advance word on the latest surveys. Colson, says Harrison, was doggedly persistent, but always polite. Some pollsters, however, were uneasy about his frequent calls. The White House was already developing a reputation for being vindictive toward the press, and they suspected that behind Colson's affability was the message "We have our eyes on you."

Colson also hoped to develop a fruitful relationship with the Gallup organization, but gave that a much lower priority than his courting of Harris, because Gallup was already thought to be sympathetic. Gallup has always been tagged as a Republican partisan—his protestations of neutrality notwithstanding—in part because he overestimated the Republican vote in each of the first four presidential elections in which he polled, and in part because many of his polling associates have worked for Republican candidates. One of his oldest associates, Claude Robinson, ran Nixon's polling operation in 1960, and Robinson's firm, Opinion Research, polled for Nixon in 1968. During that 1968 campaign, Nixon had a source within the Gallup organization who provided advance word on when the surveys were going to be taken. This allowed Nixon to time his activities so that they would have the maximum impact on Gallup's polls.

The Nixon White House thus regarded Gallup as the ideal pollster; it was his figures which set the standard for others, in particular Harris, to meet. Colson still wanted to cultivate further the relationship with the Gallup organization, but felt that he could not be the liaison. "I couldn't be the guy who held Lou Harris' hand and got him invited to things, and also do the same thing with Gallup, so I got Dwight Chapin to try to develop a personal relationship with the Gallup people so that he could go to them for in-depth information when he needed it."

According to Colson, Chapin was able to develop a pipeline for information, but it never was as fruitful as the Harris connection. In part this was a reflection on Chapin. "Dwight just isn't political. He didn't have the same sort of fingertip instincts to know what to do in his relationship with them."

Nevertheless Gallup was of some use to the administration. His son, George, Jr., was recruited for a propaganda film for the United States Information Agency. (In the film he was indentified as a "researcher respected for his objective approach," leaving the impression that it was his father speaking.) The younger Gallup testified that 77 percent of Americans stood behind President Nixon's Vietnam policies.

Like Harris, Gallup persisted in asking the wrong question on impeachment: "Do you think President Nixon should be impeached and compelled to leave the presidency or not?" When asked in late 1973 to account for the fact that the vast majority of Americans apparently believed Nixon was implicated in serious crimes, but only a minority seemed to favor impeachment, a Gallup spokesman said, "We really haven't gone into that." To Gallup's credit, however, early in the spring of 1974 he scrapped his old question and began to ask about impeachment without linking it to resignation. Immediately it became clear that a majority favored impeachment by the House in order to allow a full trial by the Senate.

Gallup also had the good sense to stay out of the White House, in contrast with Harris. "If I went, I'd be accosted by the press and they'd want to know what I promised to do for the president." Gallup, however, was not totally aloof from politics, as he did accept a presidential appointment to the United States Commission on Information, filling a slot vacated by William Buckley. Accepting that position, of course, was not nearly as compromising as polling for the White House's Domestic Council.

The administration's interest in polls was on two quite different levels. On one hand, they wanted to manipulate the published polls to document their argument that Nixon really was supported by the "silent majority," but they also wanted to get an accurate reading of public opinion for themselves. For this second purpose, the pollster on whom Colson most relied was Albert Sindlinger, who operates out of Swarthmore, Pennsylvania. Sindlinger polls on political and economic issues for private subscribers who pay up to five thousand dollars a year to get his service. He now has a syndicated column with Kevin Phillips, the former White House strategist who wrote *The Emerging Republican Majority*. In 1972 Sindlinger polled for a number of television stations. His principal work, however, is for private clients.

Most of the major pollsters—Gallup, Harris, Yankelovich—depend almost entirely on personal interviewing to gather their data. Sindlinger, by contrast, only uses the telephone. Starting at seven o'clock every night of the year, his operators dial randomly selected numbers throughout the country and run through a quick litany of questions on the economy and politics.

Sindlinger says that his system has two great advantages. First, it allows an instantaneous reading on public opinion. Second, it permits him to monitor the work of his interviewers. "Ask the other pollsters if they ever listen to their interviewers. They can't. I can monitor every damn interview my people make. I know that my questions are asked the way they ought to be, and I can hear for myself what the response is."

Colson's association with Sindlinger did not blossom until late 1971. That summer he had invited the pollster to Washington to "pick his brain" about economic issues. Sindlinger met with Colson, Haldeman and others on that visit, and in the course of discussions the subject of the 1972 campaign was raised. Sindlinger felt that the White House was being blind to how badly Nixon was being hurt by existing economic conditions, especially inflation. "Chuck and I got into one hell of an argument. So I finally said to him, 'You'd better get up to Pennsylvania and listen to my monitor and learn something about politics, because you guys are all just babes in the woods.' "

Colson had faith in his own political instincts, but he was impressed with Sindlinger, so he went that September. After much discussion, Sindlinger demonstrated his remarkable polling machine. By flicking switches on a control board, he can eavesdrop on interviews conducted with housewives in Ohio, bankers in Arizona, or widowers in Florida. From all across the country these disembodied voices report on their personal financial condition, then give their political views. At first Colson found the experience eerie, but he soon became absorbed in listening to what the people had to say. By the time the night was over, he was convinced that Sindlinger was the best pollster in the business, a view he still holds today.

After that visit, Sindlinger and Colson formed a close alliance. Colson says: "I dealt with him on a daily basis during the entire year in 1972. The great virtue of his polling is that he could tell you what had come up the night before." Colson says that the information which he got from Sindlinger was invaluable. "To me it was one of the secret weapons of Nixon's comeback. If you remember, the nadir of his presidency, to that point, was the summer of seventy-one. Everything seemed to be going wrong. Well, we just started paying closer attention to what it took to gain the public confidence that he needed." According to Colson, the decision to impose the wage-price freeze was based largely on polls which showed that it was essential that the president take forceful action to bring inflation under control.

Sindlinger's polls gave the Nixon administration the capacity to get an almost instantaneous report on the public reaction to any question or development. Sindlinger gave advice not just on what Nixon said, but on how he said it. Colson recalls, "After a speech, he'd call up and say that Nixon was using the wrong word, or that he'd hurt his credibility by claiming, let's say, that inflation had been licked." Sindlinger met privately with Nixon himself a number of times to give him a firsthand analysis of his strengths and weaknesses.

During this same period, the White House was also getting very sophisticated polling information through the Committee to Re-elect the President. Bob Teeter of Market Opinion Research oversaw the operation, but the resources of Opinion Research and Decision Making Information were also used. Pat Caddell, who was McGovern's pollster in 1972, describes the Nixon polling system with a trace of envy. "It was incredible. They had unlimited funds." All told, the committee spent at least a million dollars for political polls that year.

The choice of Teeter to do this work reflected a rift within the White House between Colson and John Mitchell. Each man wanted his own intelligence. Some in the White House saw Colson's connection with Sindlinger as an "unholy alliance," and suspected that Sindlinger was producing phony data for propaganda purposes. Though Teeter was the official campaign pollster, in many respects his work was less "political"; his operation apparently was entirely aboveboard.

Teeter's work terminated with the election, but Sindlinger continued to feed his polls to Colson and the White House well into 1974. The pollster is reticent about his role in this period, but according to Colson much of the information supplied had to do with Watergate. He says that, contrary to the conventional wisdom, Nixon had a very good reading of public opinion on Watergate from the outset, but it was impossible for the president to nip the issue in the bud by telling all. "He never could do it. Even though he knew it was hurting him, he couldn't do it. Obviously he had already had gotten his own position compromised."

John Mitchell's lawyers used a Sindlinger survey to try to convince Judge John Sirica to transfer the Watergate conspiracy case out of Washington. The poll showed that 84 percent of the District's residents believed that the defendants were guilty; only 13 percent said that they were not guilty until so proven. The venue of the trial was not changed, but Sindlinger claims that he met privately with Sirica to discuss problems of jury selection in light of the notoriety of the defendants.

It should not be surprising that from the beginning of Nixon's administration to the end, the president and his staff paid so much attention to pollsters and their polls. As a candidate in his various races, Nixon spent far more on polling than anyone else in political history. Those of his advisers, like H. R. Haldeman, whose prior experience was in advertising had been weaned on market research. Nor should it have been surprising that they would try to manipulate the polls, for they had done so before, during the 1962 gubernatorial race in California.

What is surprising, and most disturbing, is that the pollsters, who present themselves as neutral recorders of public opinion, indeed as virtual spokemen for the American people, were willing to work behind the scenes for the White House. We learned of Nixon's celebrated battle with the press because many reporters and broadcasters fought back. Nixon's parallel campaign to dominate the polls went unnoticed because the pollsters offered no resistance.

2

Polling
A Numbers Game

Public opinion polls have come to have a pervasive and often dangerous impact in America, an impact which has gone largely unrecognized and uncorrected. This book is about the polls and the men who make them. It is also an examination of the role that polls and pollsters have come to play in the political process and in our national life.

In our fascination with the tidy statistics of the polls we have become mesmerized by the measurement, losing sight of what it is that is supposed to be measured. As Lester Markel, a former editor of the *New York Times*, has said, we have confused the straws in the wind for the wind itself.

Criticism of polls and pollsters, however, is not yet in fashion. The 1972 presidential election is regarded by many as the coming of age of public opinion polling. Richard Nixon's 61 to 38 percent victory over George McGovern (1 percent went to other candidates) coincided almost exactly with the final surveys of the two best-known pollsters. In their election-eve polls George Gallup and Louis Harris had reported margins of 62 to 38 and 61 to 39 respectively.

By calling the election on the nose, the pollsters finally seemed vindicated in their long-standing claims to scientific accuracy.

Unbought copies of *How McGovern Won the Election*, which had
been published in the teeth of the polls, languished on the
shelves, reminders of just how right the polls had been.

At long last the doubts which had persisted since 1948, when
the pollsters had all but inaugurated Thomas Dewey, were qui-
eted. The few remaining skeptics were won over. The polls had
proven themselves. Though the pollsters have gone astray since
then (throughout 1975 and into 1976 Gallup and Harris gave inex-
plicably different measures of Gerald Ford's strength), their spectacular success in 1972 has thus far overshadowed more recent
failures.

To question polls now is to question "science." To worry
about the power concentrated in the hands of the pollsters is to
doubt "democracy," for we have come to regard the polls as
public opinion itself. Defending the polls, Louis Harris has written, "Anyone who says they should be done away with is really
saying that the one thing democracy can't stand is efficient journalism."

Yet the notion that the polls efficiently and reliably reflect
public opinion is a myth, a myth which in large part has been
promoted by the pollsters for their own benefit. It is a myth
which must be exploded.

Notwithstanding the success of the final presidential surveys,
1972 was in fact a poor year for the pollsters! Preliminary national
polls showed Democratic Senator Edmund Muskie to be the clear
favorite of his party for its nomination. Throughout 1971 both the
Gallup and Harris surveys had him beating the incumbent, Richard Nixon. In early 1972, just two months before the nation's first
primary, in New Hampshire, the *Boston Globe*'s poll showed that
Muskie was leading George McGovern in that state by an overwhelming 65 to 18 percent margin.

These early polls had an enormous effect on the campaign.
Muskie lined up an impressive string of endorsements from
prominent Democrats who had varying ideologies, but who
shared a conviction that Muskie's nomination—and likely his
election as well—was inevitable. Senators, congressmen, and
governors who were not necessarily that enthused about the
prospects of a Muskie presidency read the polls and decided they
would rather be part of it than be on the outside looking in. As
it turned out, of course, in hastily jumping on the Muskie band-

wagon, many of the best-known Democrats locked themselves out of their party's convention.

The battle for the Democratic nomination degenerated into a meaningless numbers game, thanks in large part to the political polls. After the *Globe*'s poll was released, McGovern aide Frank Mankiewicz announced that if Muskie were to get anything less than 65 percent of the vote in New Hampshire, it would show that the Maine senator's strength was crumbling. The Muskie people obviously did not want to be married to the 65 percent figure, but it colored their own expectations. Maria Carrier, one of Muskie's New Hampshire organizers, thought she was hedging her bets sufficiently when she said that she would "cut her throat" if her candidate did not get at least 50 percent of the vote.

The press played the same game. A week before the election, columnists Evans and Novak wrote, "At best, McGovern cannot hope for much more than 25 percent, a poor second in a drab field." The question was not whether Muskie would top the ticket but whether he would win by as much as everyone said he should. Muskie did in fact win the New Hampshire primary— which many people now forget—but because his margin of victory over McGovern was "only" 46 to 37, it was read by politicians and the press as a devastating defeat.

It was inevitable that Muskie would not do as well as the first polls indicated. Interest in primary elections is always light until the last moment. Those people who two months before the election told the *Globe*'s pollster they were leaning toward Muskie were merely saying they had heard of the senator from the neighboring state and knew nothing of his opposition. As the only visible candidate, Muskie had nowhere to go but down.

Early primary surveys are seldom more than name-recognition tests and say little about candidates' potential vote-pulling power, yet in 1972 they established a standard of performance which Muskie simply could not meet. If it had not been for the unreasonably high expectations established by the polls, Muskie's actual performance would have looked good. He had, after all, beaten a large and diverse field. His principal opponent, McGovern, had spent far more time campaigning in the state; McGovern's volunteers had canvassed the voters much more intensively.

Unquestionably Muskie's campaign was poorly run, and he

himself made some costly blunders. Nevertheless the polls were his undoing. He ultimately withdrew from the race midway through the primaries, branded a loser, yet at that time he was second only to George Wallace in total popular votes received and ahead of everyone except McGovern in elected delegates!

Muskie's rapid demise testifies to the astounding faith we place in polls: when his performance, though good, did not match the early predictions, it was concluded that it was he who had failed, not the polls. In truth, however, the early surveys, like the *Globe*'s, had been a totally unreliable guide to the actual strength of candidates.

For most of the primary campaign, Muskie's curse was McGovern's blessing. He was less known at the outset than most of his serious opponents, hence his strength was consistently underestimated. He could finish second behind Muskie in New Hampshire and trail Humphrey in Ohio, yet still be considered the winner in both races. McGovern was the unseeded challenger, the overachiever.

In California, however, the pattern was reversed. Mervin Field's final California Poll gave McGovern a massive twenty-point lead over Hubert Humphrey. No one within memory had won the Democratic primary by such a margin. The poll was taken as a sign that McGovern's strength was growing geometrically. Not only did the nomination now seem safely his, but his apparently mushrooming support made his election in November seem a real possibility for the first time. If his candidacy could catch fire among Democrats so quickly and so intensely, anything could happen.

Humphrey was stunned by the poll. He called it "the most incredible misreading of an election since the *Literary Digest* picked Herbert Hoover to beat Franklin Delano Roosevelt." In his shock, Humphrey had confused his history—it was Alf Landon who had been picked to beat Roosevelt—but Humphrey was right in saying that the poll was off. McGovern eventually beat Humphrey by a much more modest 45 to 40 percent.

Ironically, both Humphrey and McGovern are now certain Field's poll hurt their respective performances. Humphrey says, "It was like getting hit over the head with a political hammer." While he says it did not discourage him—"it just made me more determined"—it demoralized many of his workers and potential supporters. McGovern, in turn, thinks a lot of his people got overconfident, and there is some evidence to support his view. Turnout on college campuses was quite low; in earlier elections, McGovern's canvassers had been able to get out their supporters, but they may not have worked as hard in California. Pat Caddell, McGovern's pollster at the time, also notes that "the poll made Hubert the underdog, and particularly in southern California people just didn't want to see him hurt."

Whatever the net effect the poll may have had, the election left Field in the unenviable position of explaining how his figures of 46–26 could be right in the face of the actual 45–40 result. Nothing in his experience prepared him for the task. "In all my twenty-seven years of polling I've never seen the likes of the publicity that one got. The publication itself became a campaign event. Everyone quoted it, everybody vilified it."

Field's first explanation was that the poll had been its own undoing. Some people, he suggested, had voted for Humphrey just to prove the poll wrong. He also asserted that the poll could have "interfered with the electoral process" by spurring on the Humphrey workers while lulling McGovern supporters into a sense of false security.

In time Field changed his alibi, perhaps realizing that his early excuses implicated him all the more. He later invoked the old saw that the poll was accurate for the time it was taken. Humphrey had been able to pick up a great deal of support in the closing days of the election, he argued, through a massive media blitz in Los Angeles and Orange County. He also contended that Humphrey closed the gap by winning 75 percent of the undecided voters, but even if this estimate is correct, that could account for only a third of the difference between Field's projected margin and the actual result.

Throughout it all, Field seemed unable to come up with the simplest and perhaps best reason for the discrepancy, specifically, that the poll was wrong. Caddell says his private surveys for McGovern at times indicated apparent leads of as much as 15 points, but closer analysis showed that Humphrey was bound to close the gap as the election approached, particularly in southern California, where he could tap a long-standing reservoir of good will.

Although Field rightly took a great deal of criticism for his survey, the poll itself had a profound impact even after the election. Field was discredited at least for a time, but his poll remained a subconscious standard by which McGovern's performance was measured. McGovern actually won a solid victory in the most populous state in the country, yet because it did not meet the spectacular expectations established by the Field poll, it was seen in many quarters as a setback. As Caddell states, "If Field had predicted a toss-up, the story would have been that we had won four primaries in one day, crowning the whole campaign."

Instead, reporters searched for what went wrong, in much the same way they had after Muskie failed to meet the unreasonably high expectations which had been set up in New Hampshire. The conventional wisdom was that even though McGovern had captured the California delegation, the race had exposed his vulnerabilities. McGovern failed to sweep the state—he was slipping—the blue collar vote had gone to Humphrey. A seed of doubt about McGovern's appeal was planted. To some extent, it may have been self-generating.

As Humphrey himself puts it, "In less than a year's time, George went from being the beneficiary of the polls to become their victim." We are inconsistent. Most of us say we are skeptical of polls, yet in the final analysis, when an election turns out differently from the way the pollsters forecast it, we ask why the candidate went wrong!

It is ironic that McGovern's margin of victory in California was seen as a setback. Just four years earlier Robert Kennedy had beaten Eugene McCarthy by almost the same spread, but in reporting the returns before Kennedy was shot, most television commentators concluded that his victory had all but ended McCarthy's chances for the nomination, and indeed that proved to be the case.

Most of the primary polls were way off the mark in 1972. If anything, it was safest to assume that they were wrong and to bet against the front-runner. The same has been true in earlier campaigns. In the 1964 Republican presidential primaries, Louis Harris picked Rockefeller in New Hampshire, but Henry Cabot Lodge won. In Oregon, he reported in his final poll, "Ambassador Lodge appears assured of victory over five opponents." Instead of beating Rockefeller by a three-to-two margin, as Harris had indicated, however, Lodge lost by six points. Less than a week before the California primary, he reported that that state was going for Rockefeller. Harris saved himself from striking out in each of the three most important primaries, however, by announcing on the day of the election that the race had become a toss-up, with a slight edge perhaps to Goldwater. Goldwater did in fact win, but that hardly salvaged what was otherwise a disastrous primary season for Harris.

The record of the pollsters in the 1968 Democratic primaries was not much better. None of them detected McCarthy's strength in New Hampshire; he got four times the vote that the Roper survey had indicated. The polls did much better in Indiana, where Robert Kennedy won, but the pollsters fell on their face again in Oregon, where the favored Kennedy was upset by McCarthy. In California most polls correctly picked Kennedy, but overestimated his vote.

The record of published polls in other countries has, if anything, been even worse. British Prime Minister Harold Wilson called for a parliamentary election in 1970 in large part because of public opinion polls which showed that he and his Labour Party were immensely popular. Labour ran a safe and complacent campaign. On the eve of the election, almost all the major British polls still reported Labour in the lead. The Harris poll gave Labour a two-point edge, Gallup seven.

When the votes had been counted, however, the Conservatives had won by two points and had gained a majority in Parliament. Wilson was out as prime minister and Edward Heath was in. When Wilson, himself a statistician by training, was asked about the public opinion polls, he understated, "They have not distinguished themselves."

Four years later, when the British voted again, the polls were just as misleading. This time most of the pollsters correctly saw that the Conservatives would have a slight edge in the total vote, but most erroneously concluded that that meant they would retain control of Parliament. Neither majority party was able to win a clear majority, and Labour ended up with more seats. The head of Opinion Research Centre apologized in *The Times:* "Once again the opinion polls as a whole appear to have misled commentators, the public and the politicians themselves."

Similarly, the major polls have been off in Canada, where in 1972 none foresaw that Prime Minister Pierre Trudeau's Liberals would fail to command an absolute majority, and again in 1974, when none reported that he would succeed.

Given this abysmal record at home and abroad, it seems perplexing that we again and again turn to polls, not merely out of curiosity to discover who will win, but after the election, to find out what the results mean. We have made the polls the measure of success or failure. Candidates now run more against the polls than they do against one another.

Our memories apparently are short. No matter how badly the pollsters do in the early trial heats, in the primaries, and in elections in other countries, they always seem to redeem themselves in presidential elections, which are the most dramatic contests.

Experience has taught the pollsters not to call their pre-election surveys "forecasts"; instead they describe them as mere "snapshots" of opinion as it existed at the time the poll was taken. If Truman's upset victory in 1948 taught the pollsters anything it was to bury warnings in their columns which they can pull out in self-defense should the election prove them wrong. In Gallup's final pre-election poll in 1972, for example, in which he reported an overwhelming 62 to 38 percent margin for Nixon, Gallup was still cautious enough to plant the following statement nine paragraphs deep in his column: "Events subsequent to the completion of interviewing on noon Saturday could obviously affect the vote registered at the polls on election day." The final vote happened to conform to Gallup's estimate, so he did not have to use the alibi, but it was there just in case it was needed.

Before elections, pollsters hedge their bets, but after the fact they brag about how good their predictions were. Gallup, for example, claims with pride that over the past twenty-five years his polls have averaged less than three points off the mark. Harris likewise will scold readers for regarding his surveys as predictions and in the next breath claim that his "election forecasts" in 1972 were more "accurate" than Gallup's. When the polls happen to coincide with the results of an election, they are forecasts. When they are off, they are snapshots which allegedly were accurate for the time they were taken.

Gallup, Harris, and their competitors use presidential polls to pump up their reputations. Harris says: "Let me be blunt about this. The only reason I do election forecasts is that I feel the rest of society, particularly people in the seats of power, have the right to put us in the bull's-eye every four years. This is sort of our penance for being able to report week in and week out on what is happening in the country."

In essence the pollsters are saying that if they are only one or two points off in calling a national election, then all the rest of their work is similarly accurate and deserves our unquestioning respect. Of course their history in presidential primaries shows that this is not so, but unfortunately that record is quickly forgotten.

Why is it then that pollsters seem to do so well with one kind of election and so poorly with others? Daniel Yankelovich, who polls for the *New York Times* and *Time,* warns that surveys taken a day or two before a final election are unique and do not necessarily prove the reliability of any other kinds of polls. "The voter has done all the work for you. He has spent the months agonizing about the issues, resolving his uncertainties, because he has to come to a decision one way or the other by a fixed date. All the poll-taker has to do is come by and record that decision." This is particularly true for presidential elections, which get the most media attention and generate the biggest voter turnout.

By contrast, in primaries, where a vote is not considered as important, or in early trial heats, when the candidates have not become fully known, people are far less certain of their preferences for the simple reason that they have given them little thought. The pollster must try to design questions that will indicate which way such people are leaning and indeed whether they will vote. With much less to go on, pollsters must rely more on their own judgment about what *they* think people think, and, not surprisingly, they frequently go wrong.

There is a second reason for the pollsters' better performance in surveys right before presidential elections. They know that their reputations, and as a consequence their profits, depend on how close they can come to the actual outcome. When a pollster takes a survey a month or two before an election and the results do not jibe with the final vote, he can simply say that people changed their minds. This alibi is available regardless of how wrong his poll is. Thus it does not really matter whether he does his early polls carefully, for there is no way that they can be proven wrong. In September, Gallup can report that the Democrat leads the Republican by ten points or the exact reverse. Only the election-eve poll counts.

Gallup and other pollsters claim they use "generally the same methods" for their final polls as they do the rest of the year, but that simply is not so. Most of Gallup's published surveys are based on national samples of about fifteen hundred people, but for his election-eve poll in 1972 his organization interviewed thirty-five hundred. Harris normally uses samples of twelve hundred or so, but his 1972 pre-election poll was based on 3,034 "likely voters." Those people who were not registered or did not seem likely to vote were not included in that figure, so many more people were actually interviewed.

Interviewing is costly. The pollsters would not more than double their samples if they did not think the added expense was necessary. The pollsters usually are patronizing to skeptics who question how fifteen hundred people can speak for seventy-eight million voters, yet the pollsters show less than blind faith in probability by doubling their samples when their livelihood is on the line.

If Gallup and Harris make this extra effort in interviewing, for their final presidential polls, they must tighten up their other procedures as well, notwithstanding their statements to the contrary. Thus, rather than a test of general competence, election-eve polls are special events from which we cannot generalize. Most of us have learned to ignore the claims of the automobile makers who use specially tuned cars to demonstrate gas mileage and horsepower, but we accept without question souped-up surveys from those who are in the business of selling polls.

There is still another reason we must discount the pollsters' record in calling presidential elections: the simple truth is that it is not that hard to achieve. Like most parlor tricks, it is impressive until you see how it is done.

The biggest share of the popular vote ever received was Lyndon Johnson's 61.3 percent in 1964, just topping Nixon's 1972 performance. Most elections are much closer. It is hard, therefore, for any prophet to be far wrong. If you had consistently predicted that each election since 1948 would be a tie, your average error would be only six points! If you were just a little bolder and were willing to give the incumbent an edge of several points, the average error would be cut down to close to that of the pollsters. In sum, over the long run, educated guesswork could do about as well as the polls.

Since 1948 (which the pollsters understandably do not like to count), there have been six presidential elections. Three of them resulted in outright landslides for the incumbent presidents: Eisenhower over Stevenson in 1956, Johnson over Goldwater in 1964, and Nixon over McGovern in 1972. It certainly did not take a clairvoyant to see which way any of those races was going to go. Only three of the six elections can be called contests in any sense, and in one of them, the 1952 race, Eisenhower won by more than ten points.

George Gallup complains, "What really gripes the hell out of me is that nobody looks at the record," but a quick review of recent elections demonstrates how undistinguished the pollsters' record really has been. Gallup, perhaps still shell-shocked from predicting a Dewey victory in 1948, overcompensated in 1952. His raw figures showed that of those who expressed a preference, Eisenhower had a clear lead, but he hesitated calling a Republican victory because there were enough undecided voters to make the election close should most of them go for Stevenson. Instead of simply reporting the results as so many for Eisenhower, so many for Stevenson, with the rest undecided, and leaving it at that, Gallup made the mistake of imposing his own judgment on how the undecideds would vote.

He decided that because the typical undecided voter sided with the Democrats on most of the issues, the undecideds would break heavily for Stevenson. As a consequence, Gallup publicly declared that the "race was too close to call." As it turned out, Eisenhower's victory margin was much larger than average. If 1952 was really too close to call, then polls can be trusted only in landslides.

In retrospect, Gallup's error in judgment was clear. In analyzing the undecided voters' stands on the issues, he had given each issue equal weight. Although the undecided voters may have sided with the Democrats on various domestic issues, they went the other way on the Korean War, and, as it happened, that issue overrode all the others.

The percentages which are printed in banner headlines appear precise and scientific but in fact are based on a great deal of personal judgment. The pollster must decide who is likely to vote and who is not; he must determine which way people are leaning; and he must allocate the undecideds to one candidate or the other. In all of these matters there is more guesswork than science. Pollsters can easily get burned, as did Gallup.

Eisenhower's landslide re-election made it easy for the pollsters in 1956, but faced with a real contest in 1960 they came up short again. If there was ever an election which should have been called a toss-up, it was 1960's, the closest in American history, but Gallup did not hesitate to pick a winner. Luckily for Gallup, he picked Kennedy. Poor Elmo Roper was publicly embarrassed by predicting a Nixon win, but he actually came closer to predicting Kennedy's vote than did Gallup.

In 1964 Lyndon Johnson, with Goldwater's help, made it impossible for the pollsters to pick the wrong man, but Gallup and Harris both overestimated Johnson's victory margin. In 1968 it was Harris' turn to make a blunder, and he made a bad one.

The race between Richard Nixon and Hubert Humphrey appeared to be tightening, so Harris polled intensively the last few days before the election. On the basis of interviewing done on the Friday and Saturday before voting, Harris reported that Nixon held a 42 to 40 lead over Humphrey, with Wallace getting 12 percent; the rest were undecided. The Nixon-Humphrey standings were identical to those reported by the Gallup Poll.

Gallup allocated the undecided vote so that Nixon was at 43 percent, Humphrey at 42, and Wallace at 15. Gallup stood pat with these figures, but Harris continued polling. If Harris had simply added the Sunday interviews to those he had conducted the two days before, he would have correctly forecast a virtual tie at 43–43 for the two major-party candidates.

But Harris did not do that; instead, he released the results of the Sunday interviewing as if they constituted a separate poll. These results showed Humphrey ahead 43 to 40, with 13 percent for Wallace and 4 percent undecided. In his press release, Harris further noted, "If the undecideds are allocated by voter disposition, Humphrey would lead Nixon 45 to 41, with Wallace at 14."

Published only twenty-four hours before voting was to commence, Harris' report that Humphrey had overtaken Nixon had a dramatic effect. It was front-page news in many papers. The Republicans naturally jumped all over Harris, accusing him of being partisan, but he responded that he was simply carrying out his promise to report any last-minute changes.

Nixon eventually won, 43.4 to 42.7, a fact which makes Harris defensive about his projected four-point margin for Humphrey. Harris says: "It was a tough decision. If you took all three days together, it was 43 to 43, which proved to be precisely accurate. And you can imagine that I was chagrined about that." He claims, however, to have no regrets about the way in which he interpreted the data. "It's a matter of judgment—is there a surge here for Humphrey? He had been gaining, closing ground all the way through. And I said we're just going to go with our last day's polling here."

Pollsters rarely are so candid about how much personal judgment actually goes into their polls. There were some suspicions that Harris' judgement was distorted by his own sympathies for the Democrats. Not only had he long been associated with Democratic candidates, he was at that very time working behind the scenes with the Johnson administration in attempts to settle the war in Vietnam. Only Harris himself knows whether he deliberately chose to come out with a pro-Humphrey poll; even he cannot know whether his subconscious loyalties colored his vision.

Harris' denials notwithstanding, there is some evidence that he was intentionally trying to favor the Democrats. Even after he had released the survey which projected a 45 to 41 lead for Humphrey, Harris continued to poll. His final survey showed that Nixon was back in the lead. These results, however, were never made public!

Harris later told Theodore White that this last survey was taken only for "scholarly purposes" and was conducted too late to be published. That simply does not square with Harris' public pledge to report any last-minute changes. Nor is it consistent with his behavior in other elections. As mentioned earlier, on the day of the 1964 California primary Harris went on the air to announce that his earlier polls which had shown Rockefeller ahead had been superseded by an eleventh-hour survey showing Goldwater with a slight lead. Harris' 1968 polls were conducted by telephone. The results were immediately available. He could have gotten them out if he wanted to.

The whole incident demonstrates just how much subjective judgment affects polling. It is that judgment which can throw a poll way off. Speaking of his 1968 experience, Harris says, "It could happen again." Indeed it could!

The pollsters tell us to look at their record, but their exhortations must be sheer bluff, for when we do, the shallowness of their achievements becomes obvious. The pollsters would have us believe that 1948 was a fluke, that they have been on track since then, but that is not the case. And presidential polls, it must be remembered, are the best things that the pollsters do!

Pollsters are far happier doing polls on issues, for on those they can be infallible: no election can ever prove them wrong. If there is a conflicting poll on the same topic, the difference can always be explained away without embarrassing either pollster. It may be claimed, for example, that the surveys were taken at different times, hence the difference shows a shift in public opinion. Alternatively, it can be argued that variations in the wording of the questions pick up different nuances of opinion.

The pollsters thus can normally pontificate about the concerns of the American people without the fear of being caught. Occasionally, however, there are clues which expose just how speculative their reports really are. In their first 1972 post-election analysis, three of the best-known published pollsters each explained the Nixon landslide. George Gallup's figures told him that the principal issue was fear for personal safety. Writing on the same day, ostensibly about the same election, Harris reported that McGovern's pledge to give a thousand-dollar grant to each citizen was his undoing. Yankelovich did not agree with either theory. According to his surveys, Nixon had essentially won the election in May when he mined the North Vietnamese harbors of Hanoi and Haiphong. McGovern's own pollster, Pat Caddell, believes that all three explanations are wrong. He contends that overriding any political issues or personal feelings about the candidates were widespread doubts about McGovern's capacity as a decision-maker.

That the major pollsters could not agree, after the fact, on what principle issue moved the American electorate shows how subjective and tentative their interpretations must be. This has not seemed to temper their self-confidence. The pollsters continue to report what issues most concern the American people, and we, it must be admitted, continue to read what they report.

In sum, the record of the pollsters in both election and issue surveys is shaky at best. The polls were not infallible in 1948 and they are not infallible today. The pitfalls of polling and the ways in which surveys can be manipulated are discussed in detail in the chapters which follow.

There is a second myth, this one also promoted by the pollsters: that polls are neutral. If public opinion surveys had no more effect on politics than do Jeane Dixon's crystal ball forecasts have on the future, then we would not have to worry about whether or not they are accurate. Unfortunately, however, the polls have a tremendous impact on our political system and other aspects of our life. That the polls can so often be wrong makes that impact all the more disturbing.

The pollsters understandably do not want to be accused of tampering with the political process. Almost to a man, for example, they deny that favorable polls create a bandwagon of support for the lucky candidate who happens to be ahead. As proof they point to the fact that in most elections, the margin between candidates narrows as the election approaches. Indeed, Goldwater did close the gap in 1964 and Humphrey almost caught Nixon in 1968.

Viewed in a broad perspective, however, 1972 does not fit this pattern. In a Gallup Poll conducted in April of that year, Nixon led McGovern by only ten points. In successive June polls that lead increased to thirteen points, then fifteen. Although Nixon's ultimate twenty-three-point margin was less than that recorded in the polls of late summer and early fall, it was indeed larger than had been shown at the outset. Over the long run Nixon expanded his lead.

Daniel Yankelovich does not choose to call this a bandwagon effect, but he does believe the polls may have influenced some people to vote for Nixon. Shortly before the election he expressed his concern that the enormous Nixon lead in all the polls could influence voters, particularly Democrats. According to Yankelovich, people who had previously voted for Kennedy, Johnson, and Stevenson would ordinarily have a great deal of trouble pulling the Republican lever, "but when they see these polls that report that Mr. Nixon has such massive appeal, it may well reassure them that it's not so terrible to switch party allegiance."

Just how many defectors were spawned by the polls can only be guessed. McGovern himself candidly admits that he is not sure how the lopsided polls affected the final vote. "It probably does discourage some people from voting for you, but on the other hand it may make the opposition so complacent that some of them don't vote. Overall, it may cancel out."

Yankelovich's analysis and McGovern's musings can only be regarded as theories, and it is usually impossible to test such theories scientifically. There is no way of knowing what would have happened if the polls for one reason or another had showed a more even race, for we cannot run another election under controlled conditions. There is persuasive evidence, however, that the published political surveys did affect voter behavior in 1972. McGovern says, "One reason the vote wasn't heavier in '72 is a lot of people figured that because the polls showed a twenty-point spread, the election wasn't going to be close."

In absolute numbers, slightly more people voted in 1972 than in 1968, but measured in proportion to registered voters, turnout was significantly down. Many people apparently sat out the election, expressing their displeasure with both candidates. In 1968 many people similarly had bemoaned having to choose among Nixon, Wallace, and Humphrey, but when the election came, 89 percent of the registered voters cast their ballots. By contrast, in 1972, only 81 percent of those who were registered bothered to vote. In 1968 the polls, in essence, warned that the race would be close and every vote would count, while in 1972 the polls told people it would not make any difference if they voted.

The 1970 New York senatorial election was one instance in which the effect of polls on voters was measured by the polls themselves. Republican Charles Goodell was challenged for his seat by Conservative James Buckley on the right and Democratic Congressman Richard Ottinger on the left. Buckley had substantial support, it was agreed, but nowhere near a majority. The question was whether Goodell and Ottinger would split the liberal vote evenly enough to allow Buckley to win with only a plurality.

The *New York Daily News* straw poll, which gets a lot of attention in the state, published a survey showing Buckley with 37 percent, Ottinger with 30, and Goodell trailing with 24. Ottinger immediately pursued a media blitz pushing the argument that he was the only electable liberal. Goodell was under a great deal of pressure to withdraw from the race to prevent a Buckley victory. He did not, but his campaign was left in a shambles.

During this same period Daniel Yankelovich's firm was reporting on the issues and the general dynamics of the campaign. It was also keeping track of the candidates' standings, but not releasing them publicly. Yankelovich explains: "We were in an ideal position to see the massive shift that took place. Democrats and others who had been for Goodell read the *Daily News* poll and said, "My God, I'm wasting my vote.' They abandoned Goodell and went to Ottinger. It was a very dramatic change." Undoubtedly there have been many instances where the influence, be it large or small, of polls on voters has gone unmeasured yet has been decisive.

In any given election, one can only speculate on the precise effect of the polls on individual voters. There can be no question, however, about other ways in which polls can affect the outcome. Though indirect, these other means are perhaps even more powerful. Yankelovich says: "From all I've seen and read over the years, and from what I've seen in our own work, polls obviously have an effect. They have a clearer, more dramatic effect on the financial people who give support to the candidates and on the campaign workers, either boosting morale or undercutting it, making the money flow or drying it up, than they do on the electorate directly."

During the 1968 campaign, Richard Nixon said, rather prophetically, "When the polls go good for me, the cash register really rings." He might have added that when the cash register rings, a professional staff can be hired and advertising time purchased. The third element in the syllogism is that staff work and heavy advertising move a candidate's ratings in the polls up still further. Any candidate will get the aid of his best friends. One who looks like a sure winner will get help from foes as well.

There is, of course, a negative corollary to this rule. Poor polls kill contributions. Without cash it is hard to get the kind of exposure that will keep the candidate from dropping even further back. It is the Catch-22 of politics.

Humphrey found this out in 1968. Trailing by as much as twenty points in the national polls in September, he was unable to raise any money. Even *Pravda* speculated about the influence of the polls: "It is as if they are saying to the voter, why are you suffering from doubts when the election has been predetermined?"

Humphrey's finances were so bad that one of his own pollsters refused to hand over the results of a survey until he had been paid by certified check. The news from Gallup and Harris finally got better, and contributions started coming in, but there just was not time left to act effectively.

Before the nomination, Humphrey had been running well in the trial heats, but the division within the Democratic Party, the convention itself, and festering bitterness about the war caused him to take an apparent nose-dive in the national surveys. As the election itself clearly showed, these polls did not reflect Humphrey's true potential, yet Humphrey still shakes his head when he recalls the devastating effect they had on his fund-raising. "It was sort of like we had a disease." Looking back, he thinks the September polls crippled the entire campaign. "They slowed us down in the take-off. The polls were like water in the gas tank; we just didn't have that forward thrust. If I could have come out of the convention a few points, maybe even a little further, behind, I think I would have won."

Losing candidates are understandably prone to look for excuses for their failure, to blame external factors for tripping them up. Humphrey's analysis, however, is not merely defensive thinking. A switch of a quarter of a million would have given him the popular vote. Illinois, Missouri, and Ohio fell to Nixon by a hair's breadth. A switch of forty-six thousand votes would have given Humphrey New York, which then had the most electoral clout. If the Humphrey campaign had been able to get off the ground just a bit earlier, our recent history might have been entirely different.

Many would-be candidates do not even get as far as Humphrey. In 1972, Senators Fred Harris and Harold Hughes dropped out before the first primary when they found out that their low ratings in the polls killed any hope of fund-raising. George McGovern had the advantage of a more sophisticated fund-raising operation, but he still had to do well in the first primary. A low vote in New Hampshire would have been taken as a confirmation of what the polls had been saying right along. Humphrey thinks that discouraging polls were one of the factors which caused his fellow Minnesotan, Walter Mondale, to quit the 1976 presidential race.

The publication of a Gallup or Harris poll can be a political event with major repercussions. In December 1975 Gallup reported that Republicans preferred Ronald Reagan over Gerald Ford by a 40 to 32 percent margin. Newspapers throughout the country headlined the results on the front page. *Newsweek*'s cover depicted a grim-looking president, with the headline, "Ford in Trouble."

Yet for all its impact, Gallup's survey was of doubtful reliability. Only several hundred voters had been been polled, so each candidate's rating was subject to a normal error range of plus or minus eight points. Moreover, Reagan's apparent support may have been substantially inflated by the fact that he had formally announced his candidacy the day before Gallup's interviewers went out to poll. In other campaigns, people have been able to get hold of Gallup's interviewing schedule. Reagan's timing may not have been just fortuitous.

In most of the press coverage there was no acknowledgment that early surveys are subject to wild fluctuation. Indeed, a Gallup Poll just a month earlier had shown Ford in the lead by

48 to 25 percent! Polls taken the December before a presidential election have never been a good guide to who will gain the nomination, let alone win the final election. In December 1971, for example, Gallup reported that, aside from Edward Kennedy, Democrats preferred Edmund Muskie over all other candidates, just as four years earlier Lyndon Johnson was said to lead all his rivals. The Muskie and Johnson candidacies were both finished by the following April. In December 1963, Gallup stated that Republicans preferred Nixon to Goldwater, but Goldwater went on to win the nomination. Similarly, less than a year before the 1960 election, Adlai Stevenson held a lead over John Kennedy.

This history should clearly demonstrate that politicians and the press should disregard these early surveys, but because they do not, such polls can easily become self-fulfilling prophecies. Because political observers concluded that the Gallup Poll had dealt a serious blow to Ford's chances, it was so. That poll, however, was a mixed blessing for Reagan. It did give his candidacy legitimacy, but it also thrust him into the spotlight. As the front-runner, he came under much closer scrutiny by the press. For example, his proposal to cut ninety billion dollars from the federal budget had been made months earlier, but received little attention until Gallup's dramatic poll. The early lead also put pressure on Reagan to win outright victories in the primaries. Ford's pollster, Robert Teeter, stated at the time, "That poll effectively blocks Reagan from claiming that a 40 percent vote for him in a given state is a moral victory."

Primary candidates must walk a polling tightrope: they want to do well enough in the polls to get endorsements and contributions, but not so well that they cannot live up to pre-race expectations.

In the same way that political polls stimulate campaign contributions, they also generate news coverage. Candidates who are less well-known of course do poorly in early polls, which are little more than name recognition tests. Yet such candidates are usually written off by the press on the ground that "they don't have a chance." Why? Just look at the polls.

New York Times columnist James Reston has complained that we have become woefully unimaginative in thinking about potential presidential candidates. Each election year it seems the same group of tired men is trotted out of the Senate to contest again for the Democratic nomination. Why, asks Reston, do we limit ourselves to such a small field? He has suggested that, among others, men like Representative John Brademas, former Secretary of Defense Cyrus Vance, or Common Cause head John Gardner are at least as worthy of consideration as are those whose names are floated by the Great Mentioner.

It is ironic that at the same time Democrats were jittery about having no candidate for the presidency they could be so short-sighted in their attempts to find one. Whether Brademas, Vance, or Gardner (a Republican) have the qualities to be a good president remains to be seen, but if any one of them did, he could be elected. With television, a candidate can become known in an instant and respected within a matter of weeks. The Democrats could nominate a Reubin Askew, Frank Church, or Peter Rodino —pick a name—and the mere fact of nomination would give the candidate legitimacy. We are living in a time when people seem to crave new leadership, but will it happen at the presidential level? Not unless politicians and the press alike can cure themselves of their fixation with early polls.

The year 1975 provided a striking example of how the national polls can stifle the candidacies of lesser-known people. Both Gallup and Harris continually reported that Senator Edward Kennedy was the leading choice among Democrats for the nomination; given that he is far better known than most of those Democrats seeking the presidency, his lead should not have been surprising.

Kennedy had firmly stated that he would not be a candidate, but the press, dazzled by his commanding lead in the polls, churned out story after story speculating on whether he would change his mind. Could he spurn a draft? Would he let the nomination fall to George Wallace, ominously running second in the same surveys? There was little space left over for examining the character and qualifications of the other candidates, all of whom had been made to look like also-rans.

In the summer of 1975, the *New York Times* warned editorially that "the polls are building up pressures that could compel the nomination of a man who for strongly felt reasons has taken himself out of the running." Local party leaders and potential delegates who were inclined toward other candidates "cannot possibly escape the feeling that they may be engaged in a futile, and even self-defeating, activity," and "potential contributors may be wary of writing checks for those who are so regularly depicted as lagging behind Mr. Kennedy."

The *Times* analysis was certainly accurate, but its scolding tone was somewhat hypocritical in the light of the fact the *Times* itself, as a subscriber to Gallup, had been playing up the very polls it now criticized. When the *Times* rhetorically asked the pollsters what more Kennedy would have to do to be taken at his word, it should also have asked the same question of its reporters.

According to Kennedy, "I asked both Gallup and Harris to drop me from the list. They were polite, but they said no. They say it's up to them to decide what's newsworthy." Kennedy agrees that his dominance in the national surveys made it hard for others to get their campaigns off the ground, but says he does not know what more he could have done to correct the situation.

Kennedy led all other Democratic contenders throughout 1975 in the Gallup Poll, but by the end of the year his standings began to drop, apparently as people began to take seriously his statement that he would not run. Yet when Kennedy slipped, he was replaced by other old faces. In Gallup's last 1975 survey, in which Kennedy supporters were asked to state a second choice, Humphrey led with 30 percent, followed by Wallace, Henry Jackson, and Birch Bayh, at 20, 10, and 5 percent respectively. Jimmy Carter, who stood at a bare 2 percent, cracked, "These early polls only reflect name recognition, and all that means is how many times a candidate has run before—and lost!"

In many respects the stage was set for a repeat of 1972, another numbers game in which the tests of success or failure are the arbitrary figures set up by the polls. The political surveys which rated Humphrey the front-runner could well have made him as vulnerable as Muskie had been four years earlier, but Humphrey wisely declared that he was not an active candidate. Pollster Burns Roper believes that whether it is 1972, 1976, or 1980, the polls, the primaries, and the press will combine to produce what he calls a "multiplier effect." An energetic candidate can saturate a small state like New Hampshire, whose population constitutes barely one-third of one percent of that of the country as a whole.

The primary results tell us little of national preferences and attitudes, yet the press plays them up. Reporters handicap the candidates like racehorses, picking favorites according to the polls. Roper says: "The guy who does better than they thought is declared the winner. It's ridiculous, but it creates a bandwagon as you move from state to state."

As a consequence in future presidential campaigns we may again see a candidate who is not well known nationally, but who has put together a good organization, capture the nomination for the very reason that the pollsters and pundits did not rate him highly enough in their early seedings! If such a candidate can survive the early hard going, the fact that his strength has been underestimated may turn decidedly to his advantage.

Polls are likely to be even more influential when no clear winner emerges from the primaries. In brokered conventions the party leadership will try to agree on the most electable nominee. For many years the memory of the McGovern debacle will remain fresh in the minds of the Democrats. Republican regulars still recall the congressional and state house losses they suffered in 1964 by nominating Barry Goldwater. Electability, then, may be a candidate's greatest virtue. The test of electability, of course, will be the polls.

3

Three Pollsters

Louis Harris is the world's most successful and influential pollster. Between 1956 and 1963 he did private political polling in 240 different campaigns. According to Harris, he had a pivotal role in most of them. "Putting it bluntly, I elected forty-five United States senators and about twenty-three governors."

Harris became known nationally when he was hired by John Kennedy in the late 1950's to do all his polling for the forthcoming presidential campaign. Harris' association with Kennedy clinched his success. He had direct access to almost all the prominent Democrats in the country, many of whom were beholden to him.

Harris is not one to belittle his own importance. He says that after the 1960 election, President Kennedy told him, "Lou, maybe next to me you've got more power than anybody else in this country." Harris did not disagree with that assessment.

Whether he really was as close as he says to the late president is open to question; nevertheless, he was able to parlay that association into a great deal of wealth and power. Not everyone who has dealt with Harris speaks well of him, but they all acknowledge that over the years he has acquired a great deal of influence both with politicians and the press. One Democrat, prominent in national politics for almost two decades, says of Harris: "Personally, I can't stand the guy. I think he is an opportunist and a charlatan. But I don't want to be quoted bad-mouthing him. He could ruin me politically."

In 1963 Harris moved into George Gallup's domain, syndicated newspaper polls, and also expanded his business surveying, a less visible but much more profitable operation. Harris no longer does election studies for candidates; nevertheless, he has kept in close touch with those in power, Democrats and Republicans alike. During the Vietnam war Harris twice relayed secret settlement proposals from the North Vietnamese to the United States government.

Harris' biweekly column now appears in hundreds of newspapers across the country. He has written several books and dozens of articles on the state of public opinion. Politicians and businessmen alike covet his counsel. He has made millions of dollars from his polls.

Although there are more than a thousand firms, large and small, which do public opinion research, the polling profession is dominated by a handful of men like Harris. In spite of his great success and far-reaching influence, however, the general public knows little if anything about him. The same is true for even his biggest rivals. George Gallup's name is synonomous with polling, but few people are familiar with his background and beliefs. To understand polls it is necessary to know the men who make them.

George Gallup is the patriarch of public opinion polling. His career spans five decades, and in each of them he has been in the spotlight. Gallup, Elmo Roper, and Archibald Crossley were the best known of the early pollsters. Roper is dead, Crossley is out of the public eye, but George Gallup is still alive and active.

Throughout much of his professional life, Gallup has been involved in controversy; in recent years he has been able to rise above it to an extent. His sons have taken over the day-to-day operation of his various polling organizations, but Gallup himself still oversees the business. His principal role, however, is that of the elder statesman of polling. If a national magazine or television network is going to do a feature on polling, they begin by talking with Gallup. When Congress considers legislation which would regulate polls, Gallup is the first pollster to testify. To the extent that the pollsters have been able to overcome their professional jealousies to form trade organizations, it has been because Gallup strongly supported such efforts.

Gallup's offices in Princeton, New Jersey, are appropriate to his station in life. Just off Nassau Street, they are comfortable but hardly ostentatious. The red brick Gallup Building looks as if it might hold the administrative offices of some college. Indeed, Gallup, affable and outgoing, could easily be cast as the secretary of alumni affairs. He was born in 1901, the same year that William McKinley was shot, but he is younger-looking and more energetic than most of his contemporaries.

Gallup came into polling by a somewhat circuitous route. His academic training was in journalism; much of his professional life was in advertising. His doctoral thesis at the University of Iowa involved developing techniques to measure newspaper readership. He later did some teaching, toyed with some other enterprises, and ended up in the early thirties with Young and Rubicam, a New York advertising firm, where he set up their copy research department to determine what kinds of advertising were reaching people.

During this same period Gallup dabbled a bit in political polling. He did his first political survey for his mother-in-law in 1932, when she was the Democratic nominee for secretary of state in Iowa. No Democrat had been elected to high office in that state since the Civil War, but she rode in on a nationwide Democratic sweep.

Gallup did not start polling professionally until 1935, but everything he did before then, the journalism and the advertising in particular, had a strong influence on his orientation as a pollster. Much of Gallup's success is attributable to his understanding of what sells newspapers, as well as his own gift for self-promotion. When Gallup began, there were no pollsters as such. He did not pursue a career; rather, he created one.

Gallup has been accused of setting up his polling firm in Princeton just so it would have a prestigious address. He denies the charge, but he was astute enough to know that the Princeton dateline was a valuable asset. Gallup has never claimed any professional connection with the university, but the mere association of names undoubtedly has helped him, particularly in his early years. His polls would not have carried nearly as much weight if they had, for example, come out of Bayonne, New Jersey, or, for that matter, New York, where they actually were edited for a number of years. Gallup wore two hats for much of his career; in addition to his work as a pollster, he stayed on with Young and Rubicam, not leaving them until 1947.

In political circles Gallup is usually thought of as a Republican. Newspaper stories commonly refer to him as one, and when Democrats do not like his surveys, they accuse him of letting his partisan feelings get in the way of his objectivity. In part Gallup's reputation is a reflection of the record of his own polls, particularly his early ones. In each of the first four presidential campaigns, he overstated the Republican candidate's strength, sometimes considerably. In recent years, however, to the extent that his election-eve polls have been off, they have just as often overrated the Democrats as the Republicans. Still, some Democrats continue to suspect a subtle pro-Republican bias, especially in the Gallup surveys conducted several months before an election; as Hubert Humphrey found out in 1968, negative polls at this stage can scare away potential campaign contributers.

Throughout his career Gallup has tried to shake his partisan image. He insists that he is purely independent. To demonstrate his neutrality, he says he has not voted since 1936. "It would be as if the referee at the next Princeton-Yale game announced loudly that he was in favor of Princeton. Everything he would do would be a little suspect."

Gallup says, "We pride ourselves on being fact-finders and scorekeepers, nothing else." Other pollsters naturally profess to be impartial, but few go to the extreme of not voting in order to prove it. Gallup recalls a dinner party several years ago at which somebody was giving him a hard time for not exercising his franchise. "They finally turned to the late Walter Lippmann, who was another guest, and asked him what he thought. Lippmann, God bless him, said that I was right and he had never voted for exactly the same reason."

When pressed on the issue, Gallup admits that he has only bypassed voting for president; he has voted in other elections, which he does not cover in his poll. For a long while, he was a registered Republican. Though it is not generally known, he has been a registered Democrat for the past few years, but he says that this is relevant only to local politics.

Gallup's refusal to vote in presidential elections and his recent registration as a Democrat may not be enough to dispel the impression that his orientation is decidedly Republican. It is a matter not just of the supposed slant of his ratings of candidates but also of the politics of some of his associates. Though Gallup himself has not done any political polling for candidates since he began his syndicated column, many people close to him have. For the most part, these people have worked for Republicans.

The late Claude Robinson, one of Gallup's closest associates, directed all of Richard Nixon's polling in 1960. Robinson had come to work for Gallup in the late thirties, and later they set up another polling firm, Gallup and Robinson. Robinson eventually established Opinion Research Corporation, also in Princeton, which for many years was the principal Republican polling firm. Contrary to what he wants people to believe, Gallup has not operated in a political vacuum. Gallup nevertheless emphasizes that although he has been invited by several presidents, he has never visited the White House, lest he be accused of playing favorites.

According to Albert Sindlinger, who worked for him in the 1940's, Gallup has not always lived up to this arm's length policy. At that time Gallup was generally thought to have almost mystical powers to divine the future. Even then he was the premier pollster. He had successfully picked the winner in three successive presidential elections. Few people noticed that he had been as many as seven points off in his forecasts.

Sindlinger says Gallup gave in to the temptation to use his political muscle. "Before the 1948 election Dewey and Gallup were on the phone constantly. Dewey was looking for a handle on public opinion and he turned to George." Sindlinger says that Gallup, in turn, relished the role of king-maker. "George wanted Dewey to win so badly, he ended up electing Truman."

Gallup admits without hesitation that he indeed was close to Dewey. "Tom was a good friend of mine, a close friend, from the time he was district attorney." Gallup also freely admits that he was in contact with Dewey throughout the 1948 campaign. This association has left Gallup wide open to the charge of favoritism, a charge which he vehemently denies.

Sindlinger just as forcefully accuses Gallup of deliberately rigging his polls to favor Dewey. "We'd set up the headlines and draft the story, and then we would go out and do the surveys to fill in the gaps. If the results squared with our story, we'd congratulate ourselves on how smart we were. But if they didn't, then the data would be adjusted, supposedly because there was something wrong with the sample."

The story in this case, of course, was that Dewey was going to be elected president. Sindlinger swears that the headlines of a Dewey landslide were written first, then the survey results were adapted to fit them. Sindlinger sensed that Gallup's efforts on Dewey's behalf were going to backfire, so he left the firm before the election. On his own, Sindlinger took independent surveys late that fall which turned up results that contradicted Gallup's. "Gallup's sample excluded people who hadn't voted before. I found that they were heavily pro-Truman, but Gallup just didn't count them."

On the other hand, Sindlinger says that he also found a lot of people who said they were not going to bother to vote, because the polls said Dewey was such a shoo-in. "Other pollsters may deny it, but if you look at the evidence it's overwhelmingly clear that polls do influence people."

If Sindlinger's accusation is true, of course, the influence of polls in 1948 was exactly the opposite of what Gallup wanted it to be. Instead of creating a bandwagon for Dewey, they made him more vulnerable. Sindlinger believes they actually cost Dewey the election by making the Republicans overconfident.

Gallup responds unequivocally to Sindlinger's charge that the 1948 polls were rigged. "He's a goddamned liar! Never once in my entire life did I talk about a political poll with Sindlinger. He's a guy who can't tell truth from falsehood." According to Gallup, Sindlinger was in the market research end of the business and had absolutely nothing to do with the newspaper poll.

In defending himself, Gallup points out that his friendship with Dewey did not prevent him from reporting the collapse of the Dewey candidacy in the spring of 1940. He also notes that he was a friend of Progressive candidate Henry Wallace, but he clearly played no favorites with him, consistently reporting Wallace's strength in the single numbers.

Whose account should be believed, Sindlinger's or Gallup's? Within the profession, Sindlinger has the reputation of being extremely outspoken. Depending on whom you talk to, he is characterized either as a maverick who has not bowed to the polling establishment or as a man who shoots from the hip. Gallup, by contrast, is almost universally respected by his colleagues. If they themselves were surveyed, most pollsters would likely accept Gallup's word on the matter.

After twenty-eight years, of course, it is hard to ascertain just what the circumstances were. The best one can do is report directly what Gallup and Sindlinger claim. At the very least, however, Gallup was indiscreet to consult constantly with one of the candidates during the height of the campaign.

Gallup was almost as big a loser in 1948 as his friend Dewey. The pollsters became national laughingstocks, and Gallup, the most famous pollster of them all, took the hardest fall. Others were graceful in their embarrassment, but Gallup was indignant. How, he sputtered, could scientific surveys be expected to take into account "bribery of voters" and "tampering with ballot boxes"?

The next year, 1949, was not an easy one for George Gallup. A number of his subscribing newspapers threatened to cancel their contracts with him. He telegraphed his apology to the 150 papers which published his column, claiming in defense, "This is the kind of close election that happens once in a generation." Some of the papers were not won over. The editors of the *Peoria Journal* telegraphed back to terminate their contract, adding, "The Gallup Poll, had it been properly evaluated, would have told us it would be such an election."

Gallup was justifiably afraid that this sort of reaction would snowball and he would have to start his business again from scratch. He recalls that "a lot of the papers were on the verge of canceling, so we urged each one of them to go out in their own community and ask their readers whether they ought to drop us or keep us. Fortunately, every place that was done, 70 or 75 percent of the people said 'keep it,' so we weren't hoisted by our own petard! The critics were ready to discard us forever, but the people weren't."

In a fairly short time Gallup's market research business picked up, and it was evident that he had weathered his embarrassment far better than had the *Literary Digest* twelve years earlier, when it had predicted that Alf Landon would beat Franklin Roosevelt. "The fact that the polls would recover was in my mind absolutely inevitable. The one thing that sustained me was the fact that no one had ever found a better system for understanding public opinion and I didn't think anyone ever would."

Though he did survive the 1948 experience, it appears to have left a lasting mark on Gallup, and probably for the better. Though still a vigorous defender of the polls, he is not as hostile to criticism as he once was. He has certainly learned to temper his predictions, perhaps too much so: witness his announcement that the 1952 election was "too close to call." Certainly this greater caution is self-serving—it protects him in the event of another upset—but it also serves to de-emphasize, if only a bit, the impact of his surveys. It also appears that he has been able to keep a greater distance between himself and political candidates, though there are several pollsters who believe he has continued to consult with Republicans over the years, even if he has not actually polled for them. Former presidential aide Dwight Chapin had a pipeline to the firm which enabled the Nixon White House to get advance word on Gallup's polls.

Gallup now presides over a network of several companies which do a variety of polling work. He keeps an eye on the American Institute of Public Opinion, which handles the Gallup Poll, but the daily work is done by his son George, Jr. There is also the Gallup Organization, which does market research and attitudinal studies. In recent years Gallup has devoted much of his time to Public Opinion Survey, which does surveys on educational issues. There is also Princeton Survey Research, which is the fieldwork arm of the various enterprises. He seems proudest of his international affiliates, polling firms which carry his name in more than a score of countries.

Although the Gallup Poll is the best known of all these activities, it produces relatively little revenue, at least not directly. Gallup says that he breaks even on it. "We've always thought of it as a quasi-public institution." This refrain of public service is one which is repeated by most of the published pollsters. They seem to think that it is a chant which, if recited often enough, will ward off all criticism. If we are not in this for the buck, so the reasoning seems to go, then we should not get any heat for what we do.

The implication that published polls are operated as a public service is nonsense. If they lose money, and it is doubtful that they do, it is money which gets back to the pollster with a fine profit. The early pollsters quite intentionally set out to make their reputations through syndicated polls; because of their familiar names, they were able to get lucrative market research contracts.

Burns Roper, for example, says that his father, the late Elmo Roper, launched the Fortune Poll in large measure for publicity reasons. "It made a modest amount of money, but its chief purpose was to get visibility for the field and for the firm. He did elections as an attention-getter, to demonstrate the validity of the technique."

Roper says that although syndicated columns may not be particularly profitable in their own right, "the indirect monetary rewards are large." A businessman who can say Gallup or Harris did a survey for him has some clout, both professionally and personally, and that fact certainly brings business to both pollsters. New polling firms starting out often try to hook up with a local paper in order to make names for themselves. If the local paper already has a pollster, the others wait around like vultures hoping he blows an election and is fired.

Once a firm gets established, however, there may be some drawbacks to doing newspaper work. Roper says, "When we were in it heavily, I know it got us typed as doing public affairs work, and we lost out on some market research as a result." Daniel Yankelovich says that not everyone in his firm, which polls for both the *New York Times* and *Time,* is enthusiastic about doing published surveys, though Yankelovich himself finds this work irresistible regardless of profitability. On balance it seems clear that if published polls were not money-makers, one way or another, there would not be so many people trying to do them.

Gallup is the best-known pollster, but his combined enterprises are not as big, in terms of revenue, as those of some of his competitors. That seems to be entirely a function of his own style and tastes. Harris, Yankelovich, and others have all initiated special subscription services in the past several years, and they have proven to be highly lucrative. Gallup says: "That's the more profitable way to go. If you'd supply fifty or a hundred companies at twenty thousand dollars a year, you'd be way ahead of where we are. But that's never appealed to us."

Gallup seems quite content to be dean of the pollsters. He is the first citizen of the profession, Mr. Polling, and he likes that just fine. Others may earn millions or have behind-the-scenes influence, but none can challenge Gallup's prominence as a public figure. He fits the role well. He is open and unpretentious. Although polling is a gossipy and inbred business, with much jealousy and backstabbing, Gallup seems to have risen above that. He is well liked by most of his colleagues.

If one is looking for a villain, then, Gallup seems like a poor candidate. That is not to say his record is completely pure. If Sindlinger is right, Gallup's conduct in 1948 was indefensible. If, as is much more likely, his pro-Dewey polls were not actually rigged, Gallup was still guilty of being much less discreet about his political ties than he claims to be.

Gallup has mellowed in many respects. The side of him which lashed out after the 1948 debacle is not apparent now. He still rhapsodizes about the value of polls in a democratic society, but he has come to see the danger of some of the more flagrant abuses of polling. He has endorsed proposed federal legislation which would require pollsters to disclose their methods publicly, and he was one of the founders of the National Council on Public Polls, an organization dedicated to educating the press to become more sophisticated in its coverage of polls.

Gallup's organizations reflect his present temperament. They seem less aggressive about pursuing business than are some of their competitors, and they are probably less innovative in their methods. Some good people have left Gallup, feeling that the chances for advancement are limited, with two Gallup sons active in the firm. Yet even if the operation is far from being the leader in terms of either revenue or technique, Gallup himself retains the position of premier pollster, having launched the profession and promoted it for so many years.

Gallup's best-known rival, Lou Harris, differs from him distinctly in background, outlook, and personality. While Gallup's experience was in journalism and advertising, Harris' was decid-

edly political. Gallup works out of a quiet college town; Harris
has a corner office on the thirty-second floor of a building in
Rockefeller Center. Gallup is the elder statesman of polling,
while Harris, a generation younger, is at the peak of his career.
Gallup is warmly regarded by most of his colleagues. Harris is
intensely disliked.

Harris got his start in the late 1940's working for Elmo Roper.
He left in 1956 to set up his own firm, apparently taking with him
several Roper clients and leaving behind some hard feelings. At
the outset most of the work which Harris did was political. He
wrote, at one point, "For this poll-taker's part, he will never
undertake to work for any candidate he believes will set back
human progress." In Harris' case this ethic usually meant working for Democrats, and he quickly moved in on the territory
which had been staked out by other pollsters, most prominently
his mentor, Elmo Roper.

Though Harris was successful earlier, his reputation and career were really made when John Kennedy hired him as his
personal pollster in his quest for the presidency. Kennedy paid
$400,000 for Harris' surveys, much more than a political pollster
had ever received before.

Harris contends that the significance of his work for Kennedy
cannot be measured merely in dollars. "I don't think any poll-
taker before or since has sat on a strategy committee. Joe
Kennedy, Bobby and Jack Kennedy, and I—we were the inner
strategy committee. So I was part of and privy to the whole
bloody campaign. The only people who got the polls were Jack
and Bobby, nobody else."

Harris' efforts received a great deal of favorable notice in both
contemporary press stories and later accounts of the 1960 campaign. No candidate before had ever used polls as Kennedy
seemed to. Computer technology was moving into politics. Not
all the reaction was good, of course. When Kennedy critics
charged him with trying to buy his way into the White House,
Exhibit A was the bill which Harris was running up.

Ironically, at the same time Harris was getting all the attention
for his surveys for Kennedy, Claude Robinson was directing an
equally intensive operation for Nixon, but Robinson worked in
virtual anonymity. In large part, this was because Nixon did not

have a real primary challenge. In the spring of 1960 political news was dominated by stories of the Kennedy machine, of which Harris appeared to be a significant part, as it rolled through one state after another. Yet Robinson was at least as important to his client as Harris was to Kennedy. If Nixon had won the November election, then it would have been Robinson, not Harris, who would have been credited with perfecting the art of political polling.

Kennedy's victory guaranteed fame and success for the "president's pollster," as Harris was later called in a cover story in *Newsweek*. Having supposedly masterminded a presidential campaign, Harris became the number one political pollster. Democrats who wanted to win came to him, and they paid top dollar for what they got. Having dominated the political field, Harris turned in 1963 to challenge Gallup by syndicating the Harris Survey. He has developed that, plus considerable commercial work, into a firm which has made him a millionaire.

Yet, to the extent that the foundation of his career rests on the work he did for Kennedy, Harris' success is built on illusion. The quality of the polls he did for Kennedy was spotty at best, and, in the end, Kennedy gave his surveys little weight.

Harris had been hired, among other things, to test the political waters for Kennedy, specifically to see which primaries should be entered and which should be avoided. Later, as campaigning went into full swing, it was planned that he would take surveys to help shape day-to-day strategy.

Harris was still in his thirties then. He was aggressive and had the kind of winning track record that appealed to the Kennedys. At the outset, John Kennedy was quite impressed with his pollster, though that may have been a reflection more of Kennedy's vanity than Harris' talents. A former Kennedy aide says: "Face it, politicians have big egos, bigger than anybody's. Harris was smart. He'd come in with these polls that showed that everybody adored Kennedy, and Kennedy ate it up."

Early on in the campaign, however, there were signs that Harris' highly touted polls were not going to be as useful as they had been cracked up to be. They led the candidate astray in Wisconsin, for example. Ben Smith, later to become United States senator from Massachusetts, was in charge of Kennedy's effort in

Wisconsin's Tenth Congressional District. By the weekend before the primary Smith was convinced that Humphrey was going to carry the area and that Kennedy had milked all he could out of it.

Both Smith and Sargent Shriver thought Kennedy should concentrate his last day of campaigning around Madison, where they felt the effort would be most productive. Harris, however, reported that Madison was already safe and that the Tenth was winnable. Smith's unscientific but highly tuned political instinct told him Harris was wrong. Yet the pollster was insistent. Kennedy campaigned hard in the Tenth on that Monday, but to no avail. Humphrey won the district easily. More distressing, Humphrey also won in Madison. According to Theodore Sorensen's account of the incident, Kennedy blamed Harris for that loss. At the time, it was regarded as a bad setback. Only after the nomination was secure were the Kennedy people able to laugh about it.

A number of people around Kennedy began to sense that Harris' surveys were not very objective. They remembered that shortly before the voting in Wisconsin, Harris had prevailed upon Kennedy to attend a reception for his interviewers, as a way of thanking them for their efforts. The interviewers, almost all women, fawned over Kennedy, treating him more like a movie star than a politician. At the time he was delighted with the response, but later he began to wonder whether their enthusiasm for him might well have carried over into their interviewing, and thus accounted for the fact that Harris' Wisconsin polls, as well as others, were overly optimistic.

Doubts about Harris' work were confirmed in West Virginia. A 1958 Harris poll had shown Kennedy very strong there, and a survey taken in December, 1959, had him ahead of Humphrey by an overwhelming 70 percent to 30 percent.

Originally, Kennedy had planned to skip West Virginia, feeling that he would be vulnerable there. Harris' polls were a substantial factor in changing his mind. As it turned out, however, this carefully calculated planning almost backfired. After both candidates began serious campaigning, Harris' surveys showed a drastic reversal—in some areas Humphrey now led 60 percent to 40. A primary which had been entered because Harris had said it would be a cakewalk suddenly threatened to untrack the entire momentum of the campaign.

Harris, already in disfavor, tried to account for the discrepancies in the surveys, but only dug himself in deeper. Kennedy aide Larry O'Brien later wrote: "Lou explained, rather lamely I thought, that at the time of the first poll most West Virginians didn't know Kennedy was a Catholic, but by the time of the second poll they were starting to find out."

In such situations, pollsters always use the alibi that their surveys were accurate for the time that they were taken; if that is so, it is the kind of accuracy a political candidate can do without. The early Kennedy lead which Harris had reported was really no more than a reflection of his greater visibility. Kennedy was the glamor candidate and held the lead in the national polls, so it should not have been surprising that months before the primary West Virginia Democrats said they were inclined to vote for him.

Harris' polls did little more than tell Kennedy where he was, if that. They certainly did not tell him where he was going. Had Harris' techniques been more sophisticated, he would have uncovered the latent threat posed by religious bigotry, but he did not. Kennedy had to marshal his most seasoned aides and conduct an intensive barnstorming and media blitz in order to pull off the West Virginia primary.

After Wisconsin and West Virginia, Harris' polls were no longer taken seriously by the Kennedys. Having invested so much in Harris, however, having built him up as a political savant, they were not about to cut him lose. Though privately they laughed behind his back, publicly they continued to praise his work, so that his polls would have clout when they were leaked to the press and other politicians.

Harris himself may never have fully realized that the Kennedys came to use him not as a strategist but as a propagandist. There were signs, however, that Harris knew he had slipped from favor. Some people on Kennedy's staff came to believe that Harris was trying to save his neck by fudging his data to make them conform to other information the candidate was getting. According to Sorensen, "Kennedy aides O'Brien and O'Donnell grew suspicious of the whole process when they began to suspect

that the county-by-county figures forecast by the poll were influenced by their own reports on local political leaders."

Harris' final forecasts followed the usual pattern of being far
too optimistic. In the November election, he predicted that Ohio,
Wisconsin, and Washington would all go Democratic. If they
had, Kennedy would have won with an electoral margin of a little
better than two to one. As it turned out, however, all three states
went for Nixon. A further shift of fewer than five thousand votes
in Illinois and twenty-five thousand votes in Texas would have
put Richard Nixon in the White House at the beginning of the
decade instead of the end. The day after his hairsbreadth victory,
Kennedy remarked to Harris, "Well, Lou, you're still in business
—just."

Political polling is intoxicating business. Even in a presidential
campaign, a pollster usually has direct access to the candidate,
while the rest of the staff must go through the manager. All of
Harris' reports went directly to John or Robert Kennedy. Many
pollsters get an overinflated view of their own importance to the
campaign, and Harris was no exception. Even today, his private
conversations and his public speeches are peppered with references to President Kennedy.

After the election, Harris had no hesitation about ballyhooing
the work he had done for Kennedy in 1960. Writing in the *New
York Times*, he said, "When polls figure largely in the outcome of
a major victory, such as . . . President Kennedy's in West Virginia
in May, 1960, the poll-taker becomes a kind of political miracle
worker." If there was any polling miracle in West Virginia, however, it was that Harris was not fired on the spot!

In fairness to Harris, it is difficult to say whether he cynically
exploited his connection with Kennedy, inflating it far beyond
what it really was, or whether he was simply swept off his feet
by the Kennedys and lost all perspective about his function.
Kennedy later told various people of how hard it had been to get
Harris to restrain himself. On one occasion during the campaign,
while Kennedy was taking a bath, Harris briefed him on his
latest survey. Harris was spinning an elaborate web of analysis
and advice, but Kennedy, who had confidence in his own political savvy, shut him off, saying: "Just give me the numbers, Lou.
I know what they mean."

Kennedy apparently liked Harris personally, but, contrary to the pollster's claims, he did not rely on him much for political advice. Theodore Sorensen characterized the relationship in a gentle but unambiguous manner: "The senator also felt that a pollster's desire to please a client and influence strategy sometimes unintentionally colored his analysis."

Harris, however, has persisted in seeing himself as a strategist, not a mere statistician. During the campaign he would sometimes engage in flights of fancy about his future after the election. Though he now claims that he never asked Kennedy for any favors, some people close to Harris at the time say he would wistfully imagine himself as director of the CIA one day and secretary of commerce the next. When he was feeling down, he would fret that he did not have the money to move in rarefied Washington circles. Another side of him was more realistic, but in the end, when no appointment was forthcoming, the Walter Mitty in him was crestfallen.

In essence Harris suffered from the same malady which had affected the work of his Wisconsin interviewers. Indeed he became such a sycophant, and his polls were so clearly biased by his adoration of Kennedy, that he eventually had to be, in the words of Shana Alexander, "eased out of the White House."

Though his relationship with Kennedy did not go as far as he might have liked, Harris certainly fell on his feet. In 1963 he initiated the Harris Survey in direct competition with the Gallup Poll, which by then had been in operation for almost thirty years. Simultaneously he stopped doing private political work; his last poll for a candidate was for former Canadian Prime Minister Lester Pearson. If Harris were to do nationally published polls, he had to swear off political work for clients. There would be an obvious conflict of interest if what he was publishing might hurt one of his clients.

Harris has not been able to shed his reputation as a Democratic partisan, however, and has been vulnerable to the charge that his polls are skewed to the left. This conclusion is based not just on his past associations, but on the record of his poll. It is true, for example, that in each of the three presidential elections in which he has published surveys, he has overestimated the Democrat's share of the vote. Just as many politicians are sure that Gallup is pro-Republican, Harris is still commonly regarded as a Democratic pollster.

It seems somewhat odd that Harris would abandon private political polling at a time when he was the undisputed leader in the field. Harris says two things prompted his decision; one seems quite plausible, but the other less so.

Harris admits that he came to feel that he was working himself to the point of physical exhaustion without getting any tangible rewards. In the period from 1956, when he set up his firm, through 1963, he was involved in some 240 elections, more than enough for twenty lifetimes. In some respects his very success was costly. Harris says he had an ironbound rule that once a person was elected, he would not accept any business from the unit of government that person was connected with. He properly felt that any other policy might involve a conflict of interest: what might be good research for the government might not be good for his former client. The more successful campaigns he was involved in, the less government work he could take on. "My problem was that I had built up all this power, but I couldn't really use it. It wasn't a healthy thing." In essence, Harris got out of political polling so he could cash in some of his chips.

Harris' second explanation for the shift, namely that he felt uncomfortable having so much political power, seems perplexing in light of the way he has continued to cultivate his influence. Harris says that he increasingly saw that his polls had enormous impact on his clients. This influence was so great that it disturbed even Harris.

He recalls, for example doing a poll in 1961 for Pat Brown, then the governor of California, which showed Brown running some thirty points behind Richard Nixon for the election the next year. According to Harris, Brown wanted to drop out of the race then and there, but Harris helped convince him that that would be premature. Harris' advice proved to be right, of course, but he found the experience disquieting. "I realized that I was really wielding a lot of power. Our surveys were having a life-or-death impact on whether a guy remains an incumbent governor, whether he runs again, whether the life of California will be changed by it."

Harris gave up the political work, with its attendant pressures and responsibilities, and turned to the much more profitable syndicated and commercial fields. As a result he became a rich man. He swears, however, that the most visible part of his polling empire, the nationally syndicated Harris Survey, is a losing proposition. "It costs us maybe three hundred and fifty thousand dollars a year to generate and we get back eighty thousand dollars. So it's not viable economically."

Harris' poor-mouthing is nonsense. It is true that newspapers do not have to pay a great deal for the rights to publish his column, so even though the eighty-thousand-dollar figure sounds low, it may be correct. There is no way, however, that it can cost him a third of a million dollars to generate material for his feature.

Questions for the Harris Survey are piggy-backed onto other surveys, so that although a typical Harris poll may contain as many as two hundred questions, only a fraction are intended for publication. Some of these deal with timely issues, others involve possible pairings for future elections and still others may be about favorite sports. Harris milks these results, releasing them a bit at a time to fill out his twice-weekly column. Thus, the public may be confused when he states publicly, as he has, that the Harris Survey is based on intensive hour-and-a-half interviews. The material for a typical column is gathered in a few minutes or less.

In coming up with his $350,000 figure, Harris is apparently counting the entire cost of conducting the surveys in which his newspaper questions are asked, yet most of those questionnaires are devoted to asking questions on behalf of private, paying clients. People who have no basis by which to judge his claim of losing money apparently swallow his contention, but several pollsters who do newspaper work say that it just does not ring true.

Pat Caddell has considered getting into newspaper polling. He calculated it would cost $300,000 a year to produce information for a biweekly column, but that is with absolutely no piggy-backing. Grafting a newspaper survey onto commercial research saves an enormous amount of money.

Harris apparently likes to cry poor for two reasons. He wants

to think of himself as being engaged in public service. By cloth-
ing himself in the robes of philanthropy, he may feel better
shielded against criticism. He may also fear that it would appear
unseemly for him to profit personally from an enterprise which
has essentially become a political institution. Whatever his motivation, his statements simply are not accurate.

In spite of his material success, Harris does not seem to have
found the contentment Gallup has. Indeed, he seems obsessed
with proving himself the better pollster, citing the flimsiest of
evidence to make that point. He takes pride, for example, in the
claim that in 1972, "our poll was the most accurate of any poll
published—Nixon had 60.8 percent and we had said 61." Gallup,
by contrast, had said 62 percent. The difference, of course, is
insignificant. Furthermore, if one judges the pollsters' accuracy
by McGovern's performance, Gallup actually did better, forecasting exactly his 38 percent support, while Harris had said 39
percent. The actual margin between Nixon and McGovern was
one point more than Harris had forecast, and one point less than
Gallup had thought. Thus, if 1972 was a contest between the two
major pollsters, the result was an absolute draw.

It is ironic that Harris feels this need to best Gallup, for most
of those in the polling profession think Harris' work is certainly
more sophisticated. Harris is especially respected, for example,
for his ability to formulate questions which go deeper than do
Gallup's. Yet Gallup is the name known to more people generally, so Gallup is on top.

Being the second best-known pollster apparently grates upon
Harris and may have caused him to make some mistakes of judgment in an effort to scoop Gallup. In 1968, Harris' polling over
the final weekend before the voting indicated a virtual tie between Nixon and Humphrey, but Harris believed he saw a trend
in the figures, and thus projected a four-point victory for Humphrey. Had he been right, of course, he would have been the hero
and Gallup the goat, but the opposite was the case.

Gallup has always avoided polling in presidential primaries.
Harris, however, has taken them on. His overall record has
confirmed Gallup's decision to leave them alone. The volatility
of primaries is such that pollsters are bound to get burned. Harris
recalls that in Oregon in 1964 he stopped polling several days

before the Republican primary. His last survey had shown Henry Cabot Lodge leading Rockefeller by an almost three-to-two margin.

Harris considered taking just one more poll, but his interviewers were tired, so he decided to stand pat. Harris got restless, however, so he went out on his own. "I remember going to a retired nurses' home and interviewing these biddies, you know, sixty-five years old and over, and my God, they were all switching to Rockefeller!" Many of them resented the fact that Lodge not only had not campaigned in the state, but had not even said that he was a candidate. "I knew in my bones that Rockefeller was going to win, but there was no way that I could legitimately get up and say it."

Harris recounts the story as an example of the need to poll up until the last minute, particularly in primaries. There is, however, an even more important lesson in the incident, one which Harris himself may not realize. Harris was willing, indeed he felt obligated, to hold his tongue and stand by a poll he knew to be wrong!

The pollster's percentages acquire a magic, a potency of their own. Once spewed out, they cannot be called back. We sometimes admire captains who stay aboard foundering ships, but at least they get their passengers and crew into lifeboats; the fact that the ship is sinking is not kept secret. By contrast, Harris and other pollsters take everyone else down with them: "My survey right or wrong."

Harris professes to be ready to do primary polling again, if somebody is willing to underwrite it. In the past, only the national newsmagazines have been able to bankroll polls in a series of primary states. In recent years, however, there have been fewer sponsors, a reflection perhaps of the problems Harris has had in the past.

If Harris still has to play second fiddle to Gallup in terms of fame, he does not have to bow to any other pollster in terms of wealth. Several years ago, he sold his firm, Louis Harris and Associates, to the brokerage house Donaldson, Lufkin & Jenrette, and profited nicely on the transaction. In 1975 Harris Associates was sold again, this time to the Gannett newspaper chain. Harris says, "I'm one of the few people who has been paid twice for selling his company." Harris was disturbed at the *Wall Street Journal* report that he had come away with something less than a million dollars in the second deal, when in fact he got $1,397,000.

Harris' firm appears to be doing well under its new ownership. Several years ago, Harris followed Yankelovich's lead and initiated a new public opinion service for those wealthy enough to afford it, principally corporations, but also a few foundations and agencies. Each of the forty clients pays twenty-five thousand dollars a year for the privilege of subscribing to the *Harris Perspective*, which is issued monthly. Harris describes it as "an early warning business, where we take up an area like privacy, product safety, or energy." In addition to polling public opinion, Harris may also survey executives in certain areas of business or communications.

The ostensible purpose of the service is to keep business and political leaders in touch with changing public attitudes. Harris says, "These people spend twenty years fighting their way to the top, and when they get there they are twenty years out of date with what is going on in the world." Harris likes to give the impression that the subscription list is limited to forty, the implication being that the lucky clients are the only ones in the know. When pressed, however, he admits that he would take on more subscribers if he could get them.

Harris says he forsook political work because he was uneasy about the influence he was wielding, but even though he no longer labors in campaigns, he is still very well connected. Unlike Gallup, Harris makes no bones about being in touch with the political leadership of both parties. His conversation is laced with references to congressmen, governors, and presidents he has known.

Though Harris is reluctant to talk about some of his behind-the-scenes activities, such as his connection with Nixon, he makes no bones about the fact that one reason he syndicates his newspaper column is to have influence with important decision-makers. "I like doing it because it's a chance to report on what the American people think and to have some impact with the movers and shakers of the world, perhaps influence their views, which I think is a healthy thing." Harris also has many connections in the business and financial worlds. For a man who says he eschews power and influence, he still seems irresistibly drawn to it.

Harris has assumed the role of the middle-man between the American people and those who run the country. He tells the powerful and the rich—his "movers and shakers"—what America really thinks. As another prominent pollster says, "Lou has come to believe that he himself is public opinion."

Harris' political experiences, his financial success, and his access to the highest powers in government and business are heady stuff, and they have left their mark on him. His demeanor is strangely contradictory. His office, though large, is not ostentatious. It is barren of books or paintings. There are no photographs of politicians. The only personal touch is a striped marlin Harris caught.

Harris is not an imposing man. Though he got his start in politics, he is not a backslapper. He seems instead like a successful accountant, one who is perhaps afraid that someone may look at his own books. Whether in private conversation or public speech, he tends to drone monotonously. He is a dull public speaker but, in an odd way, an effective one. He speaks with such authority, dropping names and numbers right and left, that audiences are often hesitant to ask him hard questions.

Harris is not well liked within the polling profession. One pollster refers to him sarcastically as "the Great One." Another says: "Harris is an egomaniac. He thinks our government consists of the president, Congress, the Supreme Court, and Lou Harris, not necessarily in that order." A third pollster, however, cautions that at least some of this hostility may be professional jealousy. There is a great deal of backstabbing in the profession, and, given Harris' success, it should not be surprising that he is the victim of much of it. In his case, however, the antagonism runs particularly deep. There seems to be a sense among the other pollsters that Harris is the Icarus of the profession, that he has overreached and gone beyond his proper orbit.

Pat Caddell represents a third generation of pollsters. Just as Harris is different from Gallup, so Caddell is different from both of them. He is above all a political pollster; anything else he does is incidental. Unlike Gallup, who was trained in journalism, and

Harris, who apprenticed for many years with Elmo Roper, Caddell is entirely self-taught. His first poll, taken as a project for a tenth-grade mathematics class, was intended to be only an exercise in probability theory, but it got him started professionally at a precocious age. While still a high school student he polled extensively for state legislators in his native Florida.

Caddell went to Harvard College, where he met two classmates who were equally fascinated with polling. The three did surveying for several prominent politicians, among them former Ohio governor John Gilligan. Shortly before their senior year, they incorporated Cambridge Survey Research and became full-time pollsters and part-time students.

Caddell's office, close to Brattle Square in Cambridge, Massachusetts, leaves an impression far different from the sense one gets from Gallup's or Harris'. It is as cluttered as Harris' is barren. The air is as transitory as Gallup's seems permanent. On Caddell's desk is a random pile of books, letters, and reports. One book is on polling, but another is titled *Why You Lose at Tennis.*

Caddell seems in good shape now, but on his desk there is a bottle of Sucaryl, a reminder of his ongoing battle with obesity. Next to the Sucaryl sits a foot-high candle in the form of a three-faced Richard Nixon seeing, hearing, and speaking no evil. On the wall behind the desk is an oil painting, in the mode of *Sports Illustrated,* of thoroughbreds pounding to the finish line. The picture is askew. There are only two photographs on the walls. Both are of George McGovern.

Caddell was introduced to Gary Hart and Frank Mankiewicz, both of McGovern's staff, in the autumn of 1971 by a mutual friend for whom Caddell had done some polling. Although the McGovern campaign had been organized for some time, it was thinly financed, so McGovern could not simply go out and hire the best pollster money could buy. Moreover, because the nationally published surveys were ranking McGovern in the single figures, as a matter of self-protection many on the staff gave political polls little weight and were not inclined to put money into an expensive polling operation.

Were it not for these circumstances, a twenty-one-year-old college student would never have ended up being responsible for all the polling of a major presidential candidate. That Caddell was willing to do the work at cost, and that he believed McGovern actually had a good chance to win the nomination—a notion which was heresy among party leaders and political pundits at the time—made him especially attractive to the McGovern people.

Caddell did a study in New Hampshire in December 1971, three months before that state's primary, the first in the nation. The survey showed, in Caddell's words, that "Muskie was going down the tubes, particularly among blue-collar workers." Shortly after taking the survey, Caddell happened to have dinner with Bill Hamilton, Muskie's pollster, and others in the Muskie campaign. He left the party giddy with the realization that the Muskie people did not sense that their man was vulnerable. Caddell thinks that if they had recognized the problem at that stage, they might have been able to deal with it.

Caddell was determined that his client would not be similarly myopic. He says, "We spent a lot of time early in the campaign explaining to people from the senator on down just what polls can and cannot do." Ranking so low in the national polls, McGovern himself was skeptical about the usefulness of surveys, but Caddell says that he soon came around.

Caddell thinks the most significant thing about the polls he was taking in 1972 was not the raw data he was uncovering, but the way in which it was used. The principal goal was to discover the candidate's potential. "There are pollsters who say that the polls are only a time frame, a snapshot at a given moment. That's a cop-out. It's a convenient alibi for mediocre polling."

Caddell says that "any moron" can draw a reasonably good sample. The hard part of polling is analysis, and it is in this area that he thinks much of the work done by his competitors is shoddy. He calls Lou Harris' polling for John Kennedy in 1960 "crude." Harris' early West Virginia surveys had shown apparent Kennedy strength without revealing his potential vulnerability as a Catholic.

If a poll is to be of any use to a political candidate, it must have some predictive value. Caddell insists that it can, provided that the questioning is imaginative. "When we lay out a questionnaire, we want to be able to take people at a time when they are not thinking about a campaign or an issue, and see how they respond to things which may come up in the future. In the interview we introduce new information, and see how people react. From that we can project how they are likely to deal with related issues."

Working with this approach in 1972, Caddell used polls to peer through the surface of a situation to see what was developing underneath. McGovern entered the Ohio primary not because Caddell's polls said that he was already strong there, but because they indicated that Humphrey had weaknesses which could be exploited.

As the campaign wore on, McGovern surprised almost everyone by piling up victory after victory. The news media played up the "prairie populist" and the people around him. *Time* did a glowing story on Cambridge Survey Research, and Caddell became a campaign celebrity: the kid pollster, the young wizard masterminding the campaign. Caddell admits that it was like living a dual life, sitting in college classes in the morning and being interviewed by the national press in the afternoon. The success that McGovern was enjoying made it all the more intoxicating. "By the time we got to California, it was like riding a spectacular wave. It seemed as if nothing could ever go wrong."

The euphoria, however, was short-lived. McGovern won in California, but not by the margin that the Field poll had forecast. Shrewd maneuvering at the convention maximized the number of McGovern delegates who were seated, but left some people with the impression that McGovern was just another politician. Poor management put McGovern's acceptance speech on early in the morning, long after most of America had gone to bed. The agonizing indecision about what to do with vice-presidential candidate Thomas Eagleton, and the flak that was raised by Pierre Salinger's contacts with the North Vietnamese, had left the McGovern campaign in a complete shambles by the end of the summer.

After the conventions Nixon held a lead of better than two-to-one in both the Gallup and Harris polls. Caddell's principal function was to try to counteract these dismal reports. Trailing candidates always complain about the polls, and the press usually ignores their complaints, but in winning the nomination McGovern had proven the polls wrong, and reporters were willing to hear how he was going to do it again.

At the outset Caddell tried to fashion explanations of why the national figures were deceptive—Nixon's support supposedly was shallow; the race was really closer in key states—but soon Caddell was reduced to making *ad hominem* attacks on the establishment pollsters. When asked, for example, to explain the difference between Gallup's results and his own, Caddell snapped, "I'd match my sample of thirteen thousand against Gallup's twelve hundred any day." Caddell publicly questioned the methods and competence of any pollster whose surveys did not agree with his, which left a number of people in the profession deeply resentful toward him.

One of the perils of private political polling is role confusion. You are supposedly hired to obtain objective research, gathering the bad news as well as the good, but you are also hired to help elect a candidate. If shilling for him will help, you shill. Caddell denies he was engaged in propagandizing for McGovern, and in fairness to him, it does seem that after September 1972 he did his best to lay low so that he would not be put in the position of betraying either his client or his integrity. Even so, he could not prevent his name from being used to lend some credence to the idea that McGovern still had a chance to win the election. "What happens is that everyone starts quoting you. People on the plane, people in the field. People would just invent figures to keep up morale, and attribute them to us."

Caddell is somewhat sympathetic with those on the McGovern staff who exploited his name in this way, but he is bitter about how easily the press swallowed these fabricated figures. "It really gets me. Not once did anybody in the press ever call me to confirm or deny these leaks." Caddell learned the hard way that when a politician hires a pollster, he buys not just his advice but also his name.

Harris rode to success on John Kennedy's coattails, but Caddell gained prominence while advising one of the most ill-starred campaigns in American political history. One might think that would have doomed his career, but he has prospered. One pollster rhetorically asks: "How can it be? McGovern admits that he misjudged the mood of the country. Caddell was the guy who was telling him he could get the Wallace vote. What kind of credentials are those?"

McGovern's demise and Caddell's success are not as contradictory as they might appear. After the convention, press coverage of the McGovern campaign was one long disaster story, with emphasis on the blunders and agonies of an effort which was careening toward defeat. It is an unwritten rule of newspaper journalism that every cloud must be given its silver lining. Whether it is a puppy rescued from a housetop in a flood or the kid pollster valiantly toiling away in an apparently hopeless effort, some upbeat note will be found to give an otherwise gloomy story a semblance of balance. In 1968 Ed Muskie, then the vice-presidential nominee of the Democrats, was the beneficiary of this phenomenen.

Nineteen seventy-two was a year in which public opinion polls had dominated the political scene, and as a consequence, Caddell was bound to be center-stage. That he was still a college student made him good copy. That he seemed to be a polling maverick, challenging both the surveys and the assumptions of establishment pollsters like Gallup and Harris, made him a sympathetic figure. All these things coalesced to make him the darling of the political press. No one took him to task, even though he was the navigator of a ship which had run hard aground.

Caddell's success rests largely on his good luck in being at the right place at the right time. In spite of his talents, if he had not been connected with the McGovern campaign, Caddell would probably still be struggling to make a go of it as a pollster. Though his progress was phenomenal, it does demonstrate that polling is an easy business to enter, providing one has clients. There is no big investment required; computer time and interviewers can be obtained as each job comes in. For every Pat Caddell, there are scores of other people his age in politics, journalism, or advertising who are waiting for the big break that will put them on the road to becoming the next George Gallup or Lou Harris.

Caddell certainly capitalized on his opportunity. Clients now seek him out, instead of the other way around. In 1972, in addition to working for McGovern, his firm handled a half-dozen congressional races. In 1974 it was involved in thirty senatorial and gubernatorial campaigns. Edward Kennedy retained Caddell to do the polling for his 1976 senatorial race. Caddell has also toyed with doing a syndicated newspaper poll, and already he has a tie-in with columnists Rowland Evans and Robert Novak, who regularly conduct informal surveys.

Political polling, however, is an uneven business. In the months before an election the demands are intense, but in off-years things go slack. Caddell claims to be uninterested in straight market research, even though other pollsters have found it most profitable. "You can only stretch yourself so far, and toothpaste isn't really my big interest."

In 1974 his firm made its first big effort to move beyond pure political polling, offering a special subscription service, *Cambridge Reports*. Similar services, sold to businesses, banks, and unions, have been big money-makers for pollsters like Yankelovich and Harris, but *Cambridge Reports* had a rather shaky start. Caddell blames this on the nervous economy, but it may actually be due to the fact that he has been typecast as a political pollster, and a partisan one at that. The venture has been an expensive one, and some people in Caddell's firm fear that it may prove to be their first real setback.

Caddell's career has been spectacular thus far, yet his future is not entirely clear. During the 1972 campaign, Caddell billed himself as a polling maverick, maintaining that his youth and relative inexperience were strengths, not weaknesses. He prided himself in having a fresh outlook, free from preconceptions which might blind him to surprises which always lurk in the data. At the time, he said that he expected to be good for "only five or ten years." Now that more than four years have passed, he admits that he may have been a little hasty in saying that any pollster over thirty was not to be trusted.

Some people who have worked closely with Caddell, however, question whether he has already lost some of his freshness. One, who asked not to be identified, says: "He gives his clients these impressive reports, but they all say just about the same thing. He conducts the poll, then plugs the results into the same basic format. It's just like a reporter cranking out wedding announcements or a lawyer doing wills."

Caddell also seems to have alienated a number of people in the press and in politics. He is still very adroit at leaking polls to prominent columnists who seem to give his work unquestioning respect, but some reporters think that in building up the "kid pollster" they have created a monster. Caddell has been accused of having an arrogance which would be unbecoming for anyone, but which is particularly hard to take from someone still in his mid-twenties.

Caddell has been playing big-league politics when most people his age are studying it in introductory government courses, and, to an extent, his success may have gone to his head. He professes to dislike Washington and its fast political life, but he did not hesitate to fly down from Cambridge so that he could have a walk-on part in the film version of *All the President's Men*. It is not an accident that Caddell has received all the press attention, while his partner John Gorman, who works at least as hard, remains anonymous.

Caddell has also gained something of a reputation for political disloyalty, polling for a candidate in one election and his opponent the next. Rather than commiting himself early to one presidential candidate for 1976, he hedged his bets by working for several.

Caddell's professional honeymoon is over. He has made enemies as well as friends. Now that he is no longer seen as a prodigy, it may be easier to see him in proper perspective. Caddell is not a polling revolutionary, but he does have the insight to see that political polls have too often been read in a mindless, mechanical way. The paradox, however, is that the creative approach he uses carries its own dangers. Though backed up with some hard data, the ultimate analysis is really a matter of Caddell's personal judgment and, as such, is necessarily subject to error. He was perceptive in spotting Muskie's weaknesses in 1972, but he went dreadfully wrong in thinking, as he did, that an effective coalition could be forged between the McGovern left and the Wallace right.

In spite of its weaknesses, Caddell's work is more sophisticated than the polling which was done a decade ago. The same is equally true, of course, for other political pollsters, like Peter Hart, who may not be as well known to the general public.

4

*President Landon,
Meet President Dewey*

When people express skepticism about the accuracy of public opinion polls, it almost always stems from doubts about sampling. The frequent comment "I've never been polled" is really an expression of doubt that a sample of fifteen hundred people can somehow speak for an entire nation.

The history of polling is a history of trial and error. Pollsters have used one method of sampling until it has come a cropper, then have switched to another. Many people remember that the polls went wrong in 1936 and again in 1948, but it is important to know also why they failed, because it could happen again.

If anything, samples for polls have been getting smaller, not larger, over the years. The *Literary Digest*, which began its famous straw poll with the 1916 presidential campaign, mailed out millions of mock ballots for each of its surveys. On the average, about a fifth of them were completed and sent back to the magazine, a high rate of return even then.

For some time the *Digest*'s record seemed sound. In the three-way race in 1924, among incumbent President Calvin Coolidge, Democrat John W. Davis, and Progressive Robert La Follette, the magazine sent out fifteen million ballots and got back close to two and a half million. These straw votes correctly indicated that Coolidge would win a majority of the popular vote; although they overstated La Follette's actual vote, they were exactly right in forecasting which states the various candidates would definitely win.

The editors of the *Digest* always said that they made no claims for "absolute accuracy," but with its continuing success in national elections, the poll came to be trusted both by politicians and the public as a reliable barometer for forthcoming elections. Its consistency in picking winners overshadowed those times when the poll was pretty far off in forecasting the actual margin of victory.

During the 1932 campaign, the *Digest* reported a strong lead for Franklin Roosevelt. Democratic national chairman James Farley happily proclaimed: "Any sane person cannot escape the implication of such a gigantic sampling of opinion as is embraced in the *Literary Digest* straw vote. I consider this conclusive evidence as to the desire of the people of this country for a change in the national government. The *Literary Digest* poll is an achievement of no little magnitude."

Four years later, however, the magazine's luck ran out. Just as they had done in the past, the editors mailed out millions of ballots to voters throughout the country. The results that poured in during the months leading up to the election showed a landslide victory for Republican Alf Landon. In its final tabulation, the *Digest* reported that out of the more than two million ballots it had received, the incumbent, Roosevelt, had polled only about 40 percent of the straw votes.

There were signs, however, that something was wrong. The same Jim Farley who had endorsed the survey in 1932 proved that politicians are simply fair-weather friends of the published polls. His prediction of a Roosevelt sweep of all but eight electoral votes seemed obviously partisan but turned out to be exactly right.

The editors were disturbed by other signs of doubt. In the final issue before the election they wrote, "Never before in our experience covering more than a quarter century in taking polls have we received so many varieties of criticism—praise from many; condemnation from many others." But how could people question their results, they asked, when they were using the same methods which had been so successful in the past?

Within a week it was apparent that both their results and their methods were erroneous; Roosevelt was re-elected by an even greater margin than in 1932. In the first post-election issue, the *Digest* editors confessed that their faces were red, but considering the magnitude of their error, they could offer no defense, merely plead for mercy. They vowed they would not abandon polling, though they admitted they were reconsidering their methods. Although the editors were willing to try again, the American public apparently was not. A victim of both the Depression and the 1936 fiasco, the *Literary Digest* folded within a year. The practice of massive samplings died with the *Digest*. The *Digest*'s experience conclusively proved that no matter how massive the sample, it will produce unreliable results if the methodology is flawed.

The *Digest*'s survey was one expensive lesson in how not to take a public opinion poll. In retrospect the errors seem obvious. The mailing lists the editors used were from directories of automobile owners and telephone subscribers. Although the same sort of lists had provided reasonably accurate results in the past, they were clearly weighted in favor of the Republicans in 1936. People prosperous enough to own cars have always tended to be somewhat more Republican than those who do not, and this was particularly true in heart of the Depression.

It is also likely that some of the people who had been on such lists in the nineteen-twenties had, through various personal misfortunes, been off them in the thirties. More than anyone else, these voters might have been inclined to switch their allegiance from the Republicans to the Democrats, yet they were not represented in the *Digest* sample. The sample was massive, but it was biased toward the affluent, and in 1936 many Americans voted along economic lines.

As if the lopsided sample were not enough, the *Digest* compounded its error through its method of tabulating the straw votes. To maintain continuing interest in the survey, the editors mailed the ballots out at various intervals during the campaign so that each week they could report a new total. But rather than treating each week's vote as a separate poll, they made the mistake of adding the most recent returns to the previous totals. In the first months, Landon appeared to be winning by an astounding margin. Later, as the campaign progressed, the weekly returns were more evenly split, but when they were added to the running total, the trend toward Roosevelt was hidden.

A third cause of the *Digest*'s downfall was its reliance on return mail. Although the rate of return in 1936 was about 15 percent, the great majority of those who received ballots did not bother to fill them out. Apparently, their political preferences differed markedly from those of the minority which did respond.

The failure of the best known and most respected poll did not signal the end of political polling, even temporarily. At the same time as the *Digest* was forecasting a Landon landslide, several men acting independently saw it otherwise. Taking what they called "scientific polls," Elmo Roper, Archibald Crossley, and George Gallup each forecast a Roosevelt victory.

Roper was closest to Roosevelt's actual vote, 60.7 percent; he had predicted 61.7. Gallup was somewhat further off with 53.8. Yet to the general public, the point was not the margin of error, but the fact that these men, using new methods, had been able to predict the winner. That the *Digest* had failed at the same time simply highlighted their accomplishments.

Roper, Crossley, and Gallup all had some experience in commercial market research. Although their methods were crude compared with those of today, they were centered around efforts to construct unbiased samples. Each man used his own variations, but all agreed that an accurate sample for a political poll had to be a miniature electorate, that is, a sample containing all the segments of the country's population in the right proportion.

Instead of the millions of ballots the *Digest* had used, Gallup worked with no more than sixty thousand interviews, a number many called incredibly small, and he quickly cut this down to ten thousand or less for an average poll. Rather than relying on sheer quantity, the new pollsters were concerned with the qualities of the people they surveyed.

Though Gallup and others tried to use the mail, the new system really required personal interviewing, both to elicit a high percentage of cooperation and to uncover information that could not be easily checked by mail. To be certain his sample was representative, a pollster had to know the party registration, the income, the sex, and other characteristics of every person polled.

Personal interviewing also had its drawbacks, however; some were apparent from the outset, while others were not. There is always the possibility that an interviewer will cheat and fill out part or all of a questionnaire in order to get the job done faster. Also, although it was not recognized at the time, many of the interviewers avoided certain neighborhoods. Out of fear or bigotry, many of them did not bother to interview in minority areas. While lower-income people, at least at that time, voted less regularly, when they did vote they tended to vote Democratic.

In spite of these problems, the new breed of pollsters seemed to do well, though in fact their record really was not much better than that of the *Literary Digest* before its downfall in 1936. In 1940, Gallup again successfully predicted a Roosevelt victory, this time cutting his margin of error to three percentage points. Two years earlier, he had been reasonably close in forecasting the national vote in congressional elections.

The success he and his colleagues enjoyed in election polls stimulated interest and renewed confidence in polls on all sorts of issues. Public opinion polling became a continual process, no longer limited to election years. Some issue polls were serious, such as those dealing with the gathering turmoil in Europe, Roosevelt's attempt to pack the Supreme Court, and his New Deal legislation, but many other surveys were trivial. The public, Gallup declared in 1938, was not happy about the selection of Vivian Leigh to play Scarlett O'Hara in *Gone with the Wind*.

The questions some pollsters asked then tell us even more about the times than the answers they collected: "Do you believe in whipping criminals?" "Should a teacher who smokes in public be fired?" "Do you favor a federal law prohibiting the lynching of Negroes in the South?" (In response to the last question, slightly more than 50 percent said they did).

The reputations of the scientific pollsters grew rapidly. By 1940 Gallup's syndicated column appeared in hundreds of newspapers, and Roper's reports were a regular feature of *Fortune*. With their growing fame came lucrative commercial business. The government started using polls in policy-making. In 1941 Mar-

shall Field underwrote the establishment of the National Opinion Research Center, a non-profit organization devoted to the
refinement of polling techniques and the collection of public
opinion.

The pollsters were able to get off the ground in times which
were hard on other businessmen. To be sure, they were not
totally immune from criticism; skeptics wondered how long their
luck would hold. Some disgruntled Democrats complained that
Gallup, thought to be sympathetic to the Republicans, consistently underestimated Roosevelt's vote. There were mutterings
that some polls were rigged in order to make certain candidates
look strong.

In fairness to Gallup, however, it should be remembered that
just as regularly as he underestimated Roosevelt's strength, he
would overrate the Democratic vote in off-year congressional
elections. This pattern was a clue that the pollsters had not yet
solved the problem of gauging voter turnout. There were also
some rumblings about the effect of polls on the political process.
There was talk of congressional action in 1937 and again in 1943,
but nothing of substance came of it.

Partly because Gallup was a born promoter—while he was
becoming known as a pollster he was also an active vice-president
of a New York advertising firm—and also because of his evangelical devotion to polling, his name came to dominate the newly
established profession. He wrote books and articles declaring
that polling was the most useful instrument ever devised by
democracy. Polls, he promised, would reveal the true will of the
people, even more accurately than elections, which, after all, can
be rigged by bosses and crooks. Lobbyists and pressure groups
would wilt in the bright, true light of the polls. Political leaders
would hear and heed the voice of the people. Polls, according to
Gallup, were the salvation of democracy.

Not everyone shared Gallup's unqualified enthusiasm, not
even all his colleagues. In 1944, Elmo Roper stated his worry that
people had come to expect too much of polls. "I think prediction
of elections is a socially useless function. Marketing research and
public opinion research have demonstrated that they are accurate
enough for all possible commercial and sociological purposes. We
should protect from harm this infant science which performs so

many socially useful functions, but which could be wrong in predicting elections, particularly in a year like this."

Roper's cautionary words were four years premature. The 1944 polls came pretty close to the mark and their success extinguished much of the doubt which had lingered since 1936. By 1948 public confidence in the new pollsters was higher than it had ever been for the *Literary Digest*. All the pollsters agreed that even though President Truman had enjoyed public support through most of 1947, his popularity had slipped steadily since then, until the late spring of 1948, when it seemed clear he was in serious trouble with the voters. After the conventions all the major polls announced that Republican nominee Thomas E. Dewey had overwhelming support throughout the country.

Dewey was conceded such a commanding lead that the October issue of *Fortune*, which had been prepared in early September, declared: "Barring a major political miracle, Governor Thomas E. Dewey will be elected the thirty-fourth president of the United States in November. So decisive are the figures given here this month that *Fortune*, and Mr. Roper, plan no further detailed reports of the change of opinion in the forthcoming presidential campaign unless some development of outstanding importance occurs." The report concluded with the forecast that "Dewey will pile up a popular majority only slightly less than that accorded Mr. Roosevelt in 1936 when he swept by the boards against Alf Landon"!

Just as in 1936, there were some signs that things were amiss, but few people noticed them. The secretary of the Senate, Leslie Biffle, left Washington in the autumn of 1948 and traveled around the country posing as a butter and egg salesman. After sitting at lunch counters and eavesdropping on conversations in bus stations, he returned to the capital and confidently reported that Truman would be easily re-elected, but because Biffle was an ardent Democrat nobody took him seriously.

As a promotion stunt, the Staley Milling Company in Kansas City operated a feed-bag poll by printing "A Vote for the Republicans" on some of their merchandise and "A Vote for the Democrats" on the rest. By September, twenty thousand bags of feed had been sold, and, astonishingly, the Democrats held a 54 to 46 percent lead. These results were so obviously out of line with the scientific polls that the company contritely halted its survey.

The national surveys of Gallup, Roper, and the rest greatly inhibited other pollsters. The *Denver Post*, for example, commissioned a survey of Colorado voters, the results of which clearly indicated the state was going Democratic. The editors were afraid to stick their necks out, however, and they fudged their own data so that the published results gave the state to Dewey! Likewise, the *Chicago Sun-Times* took a last-minute poll which showed Truman carrying Illinois by a slight margin. The paper found these results so incredible they were never published!

The national polls had a startling effect on press coverage of the campaign. They made even the best reporters myopic. After the election the distinguished *New York Times* writer Arthur Krock apologized to his editors in a published letter: "We didn't concern ourselves with the facts. We accepted the polls unconsciously." His colleague James Reston agreed that all reporters had been "too far impressed with the tidy statistics of the polls."

In retrospect, the contortions that the newspapers went through in order to make the facts fit what the polls were saying seem ludicrous. The press explained that Truman was drawing great crowds only because of his "growing entertainment value." Newsmen were so blinded by the polls that a mob of seven thousand people who had waited hours in a thunderstorm in upstate New York to see Truman were described as "merely curious" and as "a vaudeville audience." In the closing weeks of the campaign Richard Rovere sympathetically wrote, "You get the feeling that the American people would give [Truman] anything he wants, except the presidency."

The lesson of 1948 is not only that the pollsters can be disastrously wrong. It also demonstrated how beguiled the press can become by polls. In 1948 reporters were convinced that the problems which had caused the *Literary Digest*'s failure had been cured. Today they believe that all the bugs which were in polls in 1948 have been eliminated. That is a dangerous illusion, one which could again prove embarrassing for the press.

Truman made the most of his underdog position, exhorting the voters to "show up the polls." He warned people that they were being manipulated by the pollsters: "Now these Republican polls are no accident. They are part of a design to prevent a big vote, to keep you at home on November second by convincing you that it makes no difference whether you vote or not."

During the campaign, Truman gave more than 270 speeches, and in most of them he took on the pollsters. Apparently the American people liked what they heard; nevertheless, on the eve of the election countless newspapers and magazines went to press with headlines and cover stories about "President-elect" Dewey.

The pollsters likewise did not flinch. Roper disregarded the warning he himself had given four years before and refused to hedge his prediction of a Dewey landslide. Gallup smugly wrote, "We have never claimed infallibility, but next Tuesday the whole world will be able to see down to the last percentage point how good we are."

As the election results came in, it became clear that all the major polls were very wrong. Roper and Gallup squirmed in front of radio microphones waiting for a trend towards Dewey that never materialized. Gallup, who had predicted that Truman would get only 44.5 percent of the vote, an underestimate of more than 10 percent, was still closer than most of his rivals. Roper had overstated Dewey's support by more than 35 percent! Crossley had said Truman would get only fifty electoral votes, but the president won six times that many.

Overnight the pollsters fell from their lofty position as oracles to become the butt of all sorts of humor. Goodman Ace summed up the national attitude: "Everyone believes in public opinion polls. Everyone from the man in the street up to President Thomas E. Dewey." Even William Funk admitted, "Nothing malicious, mind you, but I get a very good chuckle out of this." Nobody begrudged him that. William Funk was the former editor of the *Literary Digest.*

The initial reaction of most of the pollsters was stunned silence, but George Gallup's response was bitterly angry. Bribery and rigging had cost Dewey the presidency! The polls had been right; it was the election which came out wrong. In time he regained some of his composure, but 1948 is still a sensitive subject for pollsters who were operating then. Occasionally Gallup still mutters that the polls were close to the acceptable bounds of statistical error—they were not really *that* far off. As a parting shot, pollsters usually add, "Well, at least 1948 disproved the bandwagon theory."

When the pollsters got around to doing a post-mortem, a number of glaring errors were discovered, any one of which might have been fatal. As already mentioned, many interviewers had avoided minority areas in cities which went heavily Democratic. A few pollsters had recognized that their samples under-represented Democrats, but they compounded the problem by substituting more interviews in the rural South. The South had traditionally been Democratic, but in 1948 much of it went to Dixiecrat candidate Strom Thurmond.

It was also obvious, after the fact, that the pollsters had stopped polling too soon. Roper, who was the first of the major pollsters to stop, was also the farthest off in his results. He and the others had operated on the assumption that once a person expresses an inclination to vote one way or the other, that person's mind stays made up. Post-election research indicated, however, that one out of seven voters made his or her decision in the final two weeks of the campaign. Truman's energetic campaigning and exploitation of his position as the underdog enabled him to win back many voters. People who in the summer had felt that a Republican administration might be a welcome change had decided by November that they were not going to put "that little bridegroom on a wedding cake" into the White House. The surveys had stopped too soon to catch this movement.

There was also concern that the basic concept of quota sampling might be flawed. In building his "miniature electorate" the pollster must decide what demographic characteristics, such as race, religion, or income, will be important in the election and which will not. Nineteen forty-eight made many pollsters understandably wary about relying on any method which depended so heavily on their personal judgment. As a consequence, many of them turned to random sampling, a method which is the basis of most polling today.

The theory of random sampling is simple. If, for example, you want to know what proportion of the cars in the United States are Chevrolets, you obviously do not have to count them all. A spot check in several different places ought to give you a pretty good approximation of the incidence of Chevy ownership, provided that everyone has an equal chance of being selected for the sample. The automobiles parked in driveways in Bronxville, New York, are different from those in the Bronx.

Because random sampling is based on the probability that an unbiased sample will approximate the qualities of the population as a whole, how close a given poll comes to the truth, as it were, is really a matter of chance. One randomly drawn sample will probably differ somewhat from another. If the sampling procedure is good, most surveys should be fairly close, but a few will be far off. It is quite literally the luck of the draw.

Opinion polling operates on exactly the same premise, except that the pollsters look for attitudes rather than objective facts. If a survey can tell us approximately how many Americans own Chevrolets, it should also be able to reveal what percentage of the population are Democrats, or, by extension, which candidates are favored.

The principal difficulty in public opinion polling is discovering what people really think. It is in this area that polling methods most often fail. Yet most people persist in believing that it is the size of the sample which determines a poll's reliability.

In the fall of 1960 Joseph Alsop wrote a biting criticism of Gallup's methods, and his prime target was sampling. How, he asked, could Gallup justify using a sample of only sixteen hundred people when each percentage point in the final results would represent only sixteen people. If thirty people who grudgingly said that they were leaning toward Richard Nixon in one poll said in the next that they were now leaning toward John Kennedy, the published figures would indicate that the gap between the two candidates had changed by four percentage points. Such a shift could be decisive in many campaigns. Said Alsop, "Apparently encouraged by his failures in 1948 and 1952, Gallup cut his samples drastically in size."

The Minnesota Poll once asked people whether they thought a public opinion poll could be accurate if it were based on a sample of two thousand people. Reaction was evenly divided: 47 percent said they did, 45 percent said they did not, and the remainder were undecided. Read together with other studies, the responses to the Minnesota survey do not show that a large proportion of the population doubts the accuracy of polls, but rather that many people might if they were aware that the typical national survey is based on fifteen hundred interviews.

Doubt about sampling is the one criticism which pollsters welcome, because it is so easily refuted. If, for example, it were possible to take the millions of ballots cast in a presidential election, put them in an enormous tub, and mix them thoroughly, it would be necessary to count only a couple of thousand to get a pretty good idea of which way the election would go. Depending on luck, the preferences of the sample would likely be within a few percentage points of that of the total. This can be proven both mathematically and experimentally.

Sampling is something we do all the time. If one sip of wine is displeasing, we send the whole bottle back. We do not have to take a second swallow to be sure that the entire contents are bad. Similarly, we can get a pretty good estimate of the number of words in this book simply by counting how many appear on this one page and multiplying that result by the total number of pages in the book. On a much broader scale, manufacturers commonly rely on sampling to ensure quality control, testing only a small portion of the products they make. Such examples are so familiar to us that we do not give them a second thought, yet their very familiarity shows how much we have come to trust information drawn from samples. As Robert Reinhold of the *New York Times* has written, "For those of you who still doubt, it is recommended that you ask the doctor to remove all your blood instead of just a small sample the next time you have a blood test."

For public opinion polling, the theory of random sampling is sound. In practice, however, weaknesses creep in. For random sampling to work, each citizen has to have an equal chance of being interviewed. That means that the potato farmer in Idaho has to have as great a chance of being in the sample as does Gallup's next-door neighbor in Princeton, even though the latter might be much easier to reach.

There is, however, no unified list of all the citizens of the United States from which the pollsters could randomly select names, and even if there were, such a process would produce an unworkable sample. Fifteen hundred people drawn in that manner would be scattered across the country, which would make interviewing prohibitively expensive. To eliminate some people on the ground that they were too remote, however, would destroy the randomness of the sample. Also, it is very likely that some of the people could not be tracked down or would not cooperate, and this too would corrupt the sample. As a consequence, most pollsters must make compromises in their sampling methods. Though necessary, many of these compromises can affect the accuracy of a poll.

Pollsters commonly begin constructing their samples by randomly selecting certain areas or sampling points throughout the country. The chances of any given community being selected are proportionately weighted according to its population. The bigger it is, the greater its chance of being chosen. This selection process may go on through a series of stages, with states being selected first, then counties chosen from the selected states, and so on. Each pollster uses his own technique; some are more refined than others, but the underlying principles are the same.

Once sampling points are selected, interviewers are assigned to collect opinions. There are always a great many more households within each sampling area than the number of assigned interviews. This makes it easy for the interviewer to fill her quota, but it usually also gives her leeway as to whom to select. Thus, at this level, the sampling is not truly random. It depends on whom is at home and whom the interviewer wishes to see. Some pollsters exert stronger control over their interviewers than do others, telling them at exactly which house to begin and where to procede from there. Likewise, some poll-takers are more concerned with trying to call back at those houses where no one was home. Pat Caddell's interviewers have more discretion in the houses they choose than do Lou Harris'.

A typical Harris national survey, based on twelve hundred to fifteen hundred interviews, uses about one hundred sampling points. Thus there would normally be as many as ten points in populous states like California and New York, and none in some of the smaller ones. Harris' syndicated survey never uses sampling points in Alaska or Hawaii, as it is too costly to poll in remote areas. In recent years Gallup has used anywhere from 150 to 360 sampling points to gather about the same number of interviews. As a result, he takes somewhat fewer interviews at each point.

While the pollsters profess to have great faith in probability theory, they still examine the sample to make sure it matches the general population in terms of sex, race, geographical location (rural or urban), and the like. To some extent, this can be controlled through the interviewing process, but if one group or another is under-represented, the pollsters will either weight the figures or conduct more interviews for that particular group. In a sense, the common sampling procedure is really a hybrid of pure probability sampling and the old quota methods.

Occasionally a sample will look so unrepresentative that a pollster will discard it and construct another. Reportedly, the late Elmo Roper once randomly selected a national sample for which there was only one point representing the entire West Coast, and it was in Death Valley. Moreover, for the same survey, the only sampling points in the Southwest were on the King Ranch.

Roper saw how badly the luck of the draw had treated him—as at times it must—and threw out the sample. Yet there have undoubtedly been times when chance has produced biased samples which were not conspicuous enough to catch the pollster's eye. Randomness insures that most of the samples will be fairly representative, but it does not guarantee absolute perfection.

Pollsters use their best efforts to draw good samples for their final election polls, but they are probably less careful in their day-to-day work. Several years ago an employee of Harris who had constructed some of the firm's samples conducted a wholly unauthorized experiment. Instead of selecting sampling points randomly, the employee went to an atlas and picked out a hundred cities and towns which had interesting names. There was nothing scientific about the process; it was based simply on whim. Nobody ever caught what had been done, and the sample, such as it was, was used as the basis for at least one Harris survey.

The problems of constructing a sample become particularly acute for election polls, for their purpose is to reflect the opinion not of the general public, but just of those who vote. In the United States only about two-thirds of those who are eligible to vote are actually registered, and not all those who are registered bother to vote. Interest in presidential elections has been declining, and in other contests it is common for only a minority of the eligible population to take part.

It is possible therefore that a candidate can be supported by a small but dedicated portion of the electorate and still win. If supporters of the opposition stay home in droves, it is possible for a candidate to win an election with the help of fewer than a third of those eligible to vote. This phenomenon is dramatically apparent in the primaries. In 1972 George McGovern won the Rhode Island Democratic primary with 42 percent of the votes cast, far outdistancing all his other rivals. Yet only 9 percent of those registered voted in the primary, so McGovern actually captured the state's delegates with the active support of just 3 percent of Rhode Island's Democrats. It is conceivable, of course, that if all the Democrats had turned out, McGovern still would have swept the primary, but that seems unlikely in light of the fact that in the November final, he lost Rhode Island, a state which had gone Democratic in the three previous presidential elections by votes ranging from 64 to 81 percent.

The pollsters have not really solved the problem of determining who will vote and who will not. If you ask people directly whether they plan to vote, they will almost always say yes. To say no is as unpatriotic as attacking motherhood and apple pie. To get around this, some pollsters ask people if they are registered to vote, and terminate the interview in the one case in three in which the person is not. The danger in this, however, is it excludes people who may register between the time the poll is taken and the election. Other pollsters probe further and ask people if they voted in the last election and how deeply they feel about the present one. Even here the pollsters are relying on people to tell the truth, and as we shall see in the next chapter, that is not always safe to assume.

This filtering process can be time-consuming and complicated. Many pollsters do not bother going through it for their normal pre-election polls, implementing the safeguards only when their own necks are on the line, in the final election-eve survey.

Sometimes the discrepancies between the various polls can be at least partially explained by the fact that the pollsters interview somewhat different groups. In 1972, for example, most of Gallup's polls matching McGovern and Nixon were based on interviews with "adults," a group which includes voters and non-voters alike. At the same time, Harris' polls were based on "likely voters," people who in his judgment would probably vote. Yankelovich's surveys, in turn, were based on "registered voters," which included some people who would not turn out and excluded a few who would.

In the final pre-election poll, whether a pollster uses one question or ten to filter out non-voters, it still comes down to a matter of judgment as to who is included in the sample and who is excluded. Thus, for example, the last Gallup poll before an election will be based not on all the interviews which were conducted but only on those with the people whom Gallup judges to be most likely to vote. It is entirely possible that the raw results of a survey will show one candidate ahead, but after Gallup adjusts for turnout, the apparent leader will be behind. Thus, when we read the percentages Gallup presents in his newspaper column, we are reading not the returns of a miniature election but rather an edited version of them in which some people's votes counted and others' did not.

Historically, Republicans have been more consistent voters than have Democrats. Though there have been exceptions, the general rule has been: The higher the turnout, the better the Democrats do. Thus, in many cases, by eliminating people from the sample on the ground that they are not likely to vote, the pollsters are cutting into apparent Democratic strength. Adjustment for turnout is essential, of course, if the published results of the poll are to come close to the actual vote, yet the pollsters fail to admit that this process is more judgment than science. To do so would take some of the mystery and power from the polls.

One advantage of probability sampling is it allows precise calculation of sampling error. As mentioned earlier, just as luck insures that the sample will be fairly representative on the whole, it also insures that some samples will be more representative than others. The range of possible sampling error depends on three factors—the size of the sample, the size of the population, and whether the results indicate a landslide or a close race. Increasing the sample size somewhat reduces the range of error. It takes more interviews in a big state to have the same range of error as in a poll of a small city. Where opinion is split fairly evenly, as is often the case in elections, the range of sampling error is larger than if the division were unbalanced.

There are formulae and mathematical tables which provide the sampling error for all different situations. The usual national surveys of Gallup and Harris have sampling errors of plus or minus three or four percentage points. While many pollsters include that fact in their columns, they are not always explicit about what it really means.

If a poll shows one candidate ahead 49 percent to 43, with 8 percent undecided, that is often taken as a fairly strong lead. The apparent lead may, however, simply be the product of sampling error. The leading candidate's support may be overstated by three points, and the one who is trailing may actually be three points stronger. Thus each may have 46 percent of the vote. In such an instance, the undecided voters will be critical.

It is also possible, of course, that the error works the other way, that is, the leader is even stronger than he appears. Thus, such a poll tells us only that it is possible that one candidate may be ahead by twelve points, a near-landslide margin, or the race may be a dead heat. In the event of either outcome, the pollster can properly claim his survey was "right," that is, within the bounds of normal sampling error! Thus, when a typical poll shows one candidate leading another by six or seven points, we cannot be really sure what is happening in the campaign.

Pollsters almost never mention that the range of sampling error applies in most, but not all, cases. Ninety-five percent of the time, the range of sampling error for the typical national poll will be within plus or minus three or four points. At first glance, the 95 percent figure seems to inspire confidence, but that is not necessarily so. As forcefully as the laws of statistics require that the results will fall within the range of chance error in ninety-five out of a hundred cases, so they also demand that in the remaining five the chance error will be greater than the advertised three or four points. The dogged persistence of statistics is a boon to pollsters nineteen out of twenty times, but it is a curse the twentieth.

For any given poll, then, the odds are strongly favorable that the chance sampling error will fall within the advertised range, but considering the number of pollsters operating today and the number of surveys each one takes, the odds are equally insistent that a fair amount of them are going to be very wrong.

Pollsters try to control this by checking on the demographic characteristics of those who are in the sample. But it is always possible that the sample may be perfectly representative in terms of sex, race, party registration, and the like but be skewed in terms of candidate preference.

Before the 1960 election, Gallup averaged the results of his most recent interviews with those of the preceding survey. The idea was to dampen any fluctuation caused by chance sampling errors and to reduce the size of the error should the results of any one poll fall outside the range of expected error. Nevertheless Gallup continued to describe these combined figures as "the results of my latest poll" when in fact they were somewhat dated. Only when this practice came to light and was criticized did he stop it. Some other pollsters, however, still do the same thing. In early 1976 Lou Harris released the results of what was termed "the most recent Harris Survey" matching Ford and Reagan, when in fact the figures were based on two successive polls.

There have been instances in which pollsters or newspapers have decided to withhold a survey on the ground that the results made them suspect the sampling error was unusually large. Here again it is a matter of judgment, and people can be wrong. The newspapers in 1948 which were afraid to run their own polls showing Truman ahead were overly cautious. Undoubtedly there have been cases in which people have not been cautious enough.

The notion that one out of every twenty of his polls has to be wrong gives Washington, D.C., pollster Peter Hart a chill. "I wish someone could look at all my polls and tell me which five out of a hundred were wrong and which ninety-five are correct." Unfortunately for Hart and his readers, however, no one can.

Many pollsters are sloppy, perhaps deliberately so, when they talk about sampling error. Too often they use the words "percentage points" and "percent" interchangeably, when in fact they mean quite different things. For example, the unemployment rate increased from 4 percent to 9 percent from the late 1960's to the mid-seventies. That is a 125 percent increase; the number of people out of work more than doubled in that period. It is correct to say that the rate climbed 5 *percentage points*, but that is not the same as 5 *percent*. A 5 percent increase would have brought the rate from 4 to 4.2, a barely noticeable change, while an increase of 5 points, from 4 to 9, is disastrous.

In 1948 Gallup predicted that Truman would get only 44. 5 percent of the vote, when in fact he got almost 50. Nevertheless Gallup later claimed that his error was only 5 percent. Five percentage points, yes, but five percent, no. Truman actually got 10 percent more votes than Gallup had anticipated.

This kind of double-talk always makes the pollsters look better than they really are. In 1968, for example, an Elmo Roper poll indicated that Eugene McCarthy would get 11 percent of the vote in the New Hampshire primary. McCarthy actually got 41 percent. Can it honestly be said that Roper was "only 30 percent off," when McCarthy received almost four times the predicted vote? This kind of juggling act is obviously intended to create a standard by which it is impossible for the pollsters ever to be wrong. Perhaps it deceives the gullible, but those who see through it are bound to be distrustful of other claims made by the pollsters.

Although most published polls include a statement that the overall results may be off by several points, they rarely indicate that the range of error for the subgroup figures, such as non-whites, people over sixty, and the like, are much higher. In essence, each subgroup is a mini-sample, and because it is based on a smaller number of interviews, the range of error must necessarily be higher. Thus a poll may seem to indicate that one candidate has particular strength among college students or in the South, but even if that apparent strength looks large, it may simply be sampling error.

Syndicated pollsters like Gallup and Harris often make too much of these differences among various subgroups, stating conclusions which are not warranted by the laws of statistics. Private pollsters also can be guilty of providing information which is meaningless because it is based on too few interviews. For example, Pat Caddell did a poll for a Utah Democrat which showed that among lower-income people a certain candidate's support had apparently dropped from 32 to 23 percent. The sample for this subgroup, however, was only twenty people, so the chance sampling error was about twenty points either way! Thus it was possible that in the period the two polls were taken the candidate had actually fallen from 52 to 3 percent among this constituency or had risen from 12 percent to 43. In the same report, there was another subgroup, this one of only five people, and somehow Caddell had that constituency split 52 to 48 percent!

When such percentages appear in print and are accompanied by an authoritative-sounding analysis—"the candidate is suffering serious erosion of support among lower-income voters"— they seem definitive. Only if the reader notices that they are based on tiny samples and are subject to great error either way does it become clear that they are essentially meaningless.

Perhaps the most deceptive thing about the way pollsters present the problem of sampling is also the most subtle. Sampling error is the one variable over which the pollsters can exercise a fair degree of control. It is not, however, the only place that pollsters can go wrong. Questions may be biased, the interviewers may cheat, voter turnout may be other than expected—any of more than a hundred things may go wrong. None of those possible pitfalls are reflected in sampling error. It is impossible for the pollsters to do any better than the laws of probability will let them, and, given the number and magnitude of other sources of error, it is likely that they will do much worse.

5

Interviewing
the Public

Pollsters have been able to insulate themselves from criticism by constructing a barricade of statistical and scientific jargon. "Cross-tabulations," "area random sampling," "flow-coding"—their language sounds so technical, so formidable that most of us are intimidated into believing that we are not qualified to judge whether public opinion polling is a legitimate science.

To the same end, many pollsters affect the same pretensions assumed by faith healers, astrologers, and other quacks: it is *Dr.* Gallup's American *Institute* of Public Opinion in *Princeton,* New Jersey. New polling firms spring up right and left, and each one seems to be called Survey Research Associates or some such name which conjures up images of white-jacketed computer technicians pouring over print-outs.

If polling is a science, however, it is at best a crude one. Polling theory, in fact, is not complicated, and to understand it is to know just what polls can and cannot do. Even a public opinion poll with a perfect sample and unbiased questions has inherent limitations. When the theoretical precepts are not scrupulously applied—and they seldom are—the results become all the more shaky.

Diane Bentley* does interviewing for both Louis Harris and Pat Caddell. To follow her as she makes her rounds is to peer behind the wizard's curtain and see polling as it really is. The practice is a far cry from what the pollsters would have us believe.

Diane began working as a poll-taker several years ago. Originally, she did it principally to get out of the house since her children were both well along in school. Now, however, with family money tight, the pay is an important incentive, so she takes on any polling assignment she can get. Most pollsters imply that they have their own special force of fieldworkers, but in fact many interviewers like Diane find it necessary to work for two or more polling firms.

On a Tuesday morning in June, Diane went to a working-class section of New Haven, Connecticut, where she had to complete ten interviews. Some firms tell their interviewers to start at a certain house and work in a particular direction, but this day she was simply given a map of the neighborhood and told to begin wherever she wanted.

"If I get to choose, I like to drive around the area a bit. I look for parked cars, toys in the yard. That means that it's more likely people are around and I don't have to waste time at empty houses."

Judged by Diane's standards, the neighborhood did not look very promising, but she managed to find a street where there were a few cars. The first house she tried was surrounded by a chain link fence. The door had two locks on it and a small decal which said the premises were protected by Lectronic Alarm Systems. One supposed it was a home of a middle-aged couple, with perhaps a son in the service and a married daughter living in Indiana. The occupants, one guessed, were conservative in their politics, but no one answered the door, so whatever their views, they went unrecorded.

Next door was a rather run-down two-family house. No one was home on one side. An older man answered Diane's knock on the other side and she was pleased, as it is harder to find men at home during the day. Her pleasure was brief, however, because

*Her real name is not used here, and a few identifying details have been changed to protect her anonymity.

the man spoke so little English that an interview was impossible. Diane tried to explain what she had wanted, but her attempt left him confused and her embarrassed.

Diane crossed the street to talk with a woman who was hanging up her wash. Diane introduced herself, though with the oversized button she wore on her blouse that was hardly necessary. Diane's manner is naturally friendly, and three years of work as a poll-taker has polished her ability to put people at ease. Nevertheless, the woman refused to be interviewed; she said she had too much housework to do before her daughter came home from kindergarten. Diane persisted, saying that it was important to express one's opinion and that the whole thing was strictly confidential, but the woman still said no.

Diane looked down a side street, hoping for some sign that things would get better, but the houses looked just the same as those she had left. "This is a bad start. Sometimes the first three places I'll try, I'll get three interviews, one, two, three. Then there are days like today."

In theory, every person in the United States must have an equal chance of being selected in a sample if the survey is to be reasonably representative. In practice, however, certain types of people are much easier to find than others. Housewives, retired people, and the unemployed may be found at any hour, but young working people are hard to track down. On many issues homebodies tend to have different opinions from those who are on the move.

Many pollsters are reluctant to reveal the trouble they have getting to talk to people. When questioned on this they generally say that "people are most cooperative." If pressed, however, Louis Harris admits that upwards of 20 percent of the people his interviewers contact refuse to talk, and this figure does not include those who are not at home.

The rate of refusal has gone up in recent years, a fact that leads some to speculate that Watergate and all its fallout have caused people to be suspicious of anyone asking questions. Somebody who says she is a pollster may actually be with the FBI or the Internal Revenue Service. Also, there are door-to-door salesmen who introduce themselves as poll-takers; people who have been taken by this pitch may be wary of those who make the same claim. Whatever the reason, the kind of people who do not feel

comfortable talking to strangers are likely to have different attitudes about themselves and the world they live in, yet their
opinions are never represented in public opinion polls.

There is little the pollsters can do to eliminate the problem.
Paying people to cooperate would be very expensive and might
well bias the results. Gallup and Harris have the advantage of
well-known names, and as a result their interviewers are more
likely to be trusted. Other pollsters experience even greater difficulty in getting into people's homes.

Pollsters try to control the problem by hiring interviewers
who will have the greatest chance of winning people's trust. The
major pollsters agree that the most important criterion is sex.
George Gallup states: "Almost all our interviewers are women.
And the same is true for other firms all over the world. Women
have always been better interviewers. People are really less reluctant to talk with a woman, and women are much more conscientious."

Burns Roper says that perhaps 98 percent of his firm's interviewers are female. "Access is the big thing. Women are superior
at getting cooperation from either sex. Chivalry isn't dead—men
will be more responsive to a woman than they will be for a man.
And women aren't afraid of another woman, as they might be of
a man coming to the door." Lou Harris similarly reports that
over 90 percent of his house-to-house interviewers are women.

There are other criteria for interviewers. The late Oliver
Quayle candidly stated that the best ones are "not *too* intelligent
so they will not get too curious and involve themselves in survey
analysis." Other pollsters are more guarded on this point, for
they understandably do not want to slander their own employees, but it is clear that you can be considered "overqualified"
for public opinion research.

In short, polling is a business in which women are the drones,
and men are the king bees. There are no famous female pollsters.
Gallup admits, "In a sense, we have been able to take advantage
of women because a lot of educated women don't want full-time
jobs; they just want to get out of the house." As of yet no women
have sued the major pollsters for discriminatory employment
practices.

Daniel Yankelovich's firm is one notable exception to the general pattern. He is the firm's best-known member, but many of its senior officers, including one partner, are women. Ruth Clark, who heads the Public Opinion division, says: "I don't think there is a sexist conspiracy. At least I've never encountered it if there is." To an extent, perhaps, the barriers to advancement may be put up by the pollsters' clients, not the pollsters themselves. Women pollsters are increasingly accepted by corporate and media clients, but few candidates for public office have shown any willingness to pay female poll-takers for political intelligence and advice.

One after-the-fact explanation for the pollsters' blunder in predicting that Dewey would win the election in 1948 was that many of their women interviewers stayed out of the black neighborhoods in the northern cities, where the vote went heavily to Truman. Most pollsters try to hire people who will feel comfortable in various kinds of neighborhoods, but problems still persist. Gallup is concerned that some urban people may become inaccessible. "As the crime situation actually gets worse in some cities, it becomes more and more a problem to send interviewers into areas where the crime rate is high. We are reluctant to do it, and so are our interviewers."

When Diane Bentley interviews at night, her husband usually drives her around, though the arrangement is not completely satisfactory. "While I'm doing my interview, I sometimes worry about him out there in the car." Diane has interviewed all over Connecticut, in rich areas and very poor ones. Her one real phobia is dogs. She walked by one house without hesitating; a large German shepherd was sleeping on the porch.

Diane finally succeeded at the sixth house. A woman returning from the supermarket with two large bags of groceries said she would be willing to talk for a little while. Diane did not mention that the survey would take more than an hour. If she had, the woman would probably have declined, as would most people.

The house-to-house surveys taken by the major pollsters are getting longer and longer. A typical Harris interview takes more than an hour. Those done by Cambridge Survey Research, Pat Caddell's firm, often last an hour and a half. Caddell's polls are usually commissioned by a single political client, but most of the other pollsters use one interview to ask questions for a number of subscribers. Gallup surveys commonly start with questions which will be used for his newspaper column, but then go on to matters which are being probed for one or more private businesses.

This practice of piggy-backing allows the pollsters to spread the costs of interviewing among several clients, but it also means that the interviews are long and ponderous. Harris questionnaires can run up to thirty pages with perhaps two hundred questions. Other pollsters who do personal interviewing follow similar practices. People who are willing to be interviewed get nothing but the satisfaction of knowing their responses will fractionally affect the totals spewed out by the pollster's computer.

Diane did not know who was the principal client for the survey she was conducting, but most of the questions dealt with energy problems, which made her guess that the client was either an oil company or a public utility.

Diane followed the woman into the kitchen and sat down with her clipboard to conduct the interview while the woman put away her groceries. Diane said she just wanted "your reaction to these questions—it doesn't have to be something you would stand by." The first questions asked whether there was an energy shortage, either locally or nationally, and whether there would be one in the future. The woman hesitated. "Well, I think the gas crisis was manufactured by the companies." Diane checked the boxes that said there was no energy shortage.

Diane is essentially paid according to the number of interviews she does, so she moved crisply through the questionnaire, politely but firmly insisting on specific answers to the questions she put. As the interview went on, however, the woman got increasingly impatient. "Everything you're asking me is yes or no, black or white. I just don't think that way." She felt trapped by a question which asked her to agree or disagree with the statement, "Since Henry Kissinger failed to make peace between Egypt and Israel it looks as though he is losing his touch as a peacemaker."

She said she had never liked Kissinger. Were she to take the question literally, she must answer no, for she believed he never had had a touch as a peacemaker. But that answer, as she could plainly see, would be taken as an endorsement of Kissinger. On the other hand, she could not bring herself to say yes, as that implied that she had until recently supported him.

Diane was pleasant but persistent. "Just answer the question as best you can. Choose whichever is the lesser of two evils." The woman still could not subscribe to either of the two offered alternatives. Diane eventually put her down as "not sure" when in fact the woman had a clear and strong opinion about Kissinger.

The question, typical of many in that survey, illustrates a serious problem common to much of public opinion polling: complex attitudes are artificially forced into neat little boxes—agree/disagree or favor/oppose. No analyst, no matter how sophisticated, could divine the woman's real views from the "not sure" recorded by Diane.

The pigeonhole effect is the result of the polling mechanism. If the pollster sets out to report public opinion in some unified and comprehensible fashion, he must get answers which fit into categories which can be easily tallied. Open-ended questions which ask people to state their views in their own words allow each individual to express himself accurately, but the very diversity of expression which is produced necessarily means that it cannot be tabulated into neat "for" and "against" columns.

This pigeonholing is a fundamental weakness of public opinion polls. Even if a poll is based on perfectly constructed samples, and even if the results are scrutinized by a sophisticated analyst, the results will be meaningless if the questions were simplistic. Peter Hart is one of the few in his profession even to acknowledge the problem. "Look at the newspaper polls. All the questions are 'agree/disagree.' Well, life isn't that simple. It isn't just 'excellent, good, fair, or poor.' It makes cleaner newspaper copy to make everything 'agree/disagree,' but that's an awfully simple way of looking at public attitudes."

To the extent that other pollsters have felt this sort of criticism, their usual response has simply been to ask more questions, not better ones. Each individual agree/disagree question is flawed, but most pollsters think that the total value of a survey can somehow be more than the sum of its parts. Pat Caddell, for example, prides himself on the depth of his interpretation of his surveys, but his analysis can be no better than the questions he asks. If anything, Caddell's questions tend to be even more restrictive than those of his major rivals.

Caddell is hardly alone, however, in forcing people to choose between two positions neither one of which they support. Consider some of the questions Diane Bentley was asking. "Would you like to see the U.S. become *totally* independent of all foreign sources of energy, or not?" You must answer yes or no. You cannot say that you are concerned more with dependence on Middle Eastern oil than with Canadian sources. Nor can you qualify your answer by saying that your support of energy independence depends on how much more energy will cost when it is produced solely by domestic sources. Both of those considerations would seem to be essential to any intelligent response to the question, yet the pollsters demand a simple yes or no.

Most public opinion polls are a long string of such questions. In a 1975 survey Pat Caddell conducted in Massachusetts for Edward Kennedy, more than 130 of the 165 questions dealing with issues or impressions were either agree/disagree or some similar format.

Most of us would have trouble answering these yes-or-no questions. Our attitudes are more complex, and there are often other considerations which must be brought into play. Just as a simple yes-or-no answer would not accurately reflect our own true opinions, it also fails to represent the views of the fifteen hundred people who are polled by Gallup or Harris.

A simple test for determining the meaningfulness of poll results is to read the questions to see if you yourself would be comfortable giving agree/disagree answers. If you would not, then you must discount the results of the poll, no matter how conclusive the statistics seem.

The Gallup Poll is careful to include the wording of all questions in the releases. Gallup himself says, "We regard ourselves simply as fact-finders. We publish the questions, tell how they were collected, and let the reader draw his own conclusions from the facts."

Unfortunately, not all pollsters follow this procedure. Harris usually does not supply the wording of the questions which are asked for his syndicated column, and he is sensitive about criticism on this point. "This is the big argument I have with Bud Roper and others in our field. He says that we shouldn't report a question unless we also report the wording and the tabular results. Well, all our columns are based on eight or ten questions, and there is no way to get all that information in with a word limit of eight hundred or eight hundred and fifty."

Harris notes that complete information about his published surveys is on file with the University of North Carolina and that his firm has always made it a practice to disclose wording of questions and other information to anybody who requests it. As helpful as that may be to serious scholars, it does little good for the average reader, who is left in the position of having to trust Harris' judgment in interpreting the meaning of his surveys.

In zeroing in on simple agree/disagree questions, the pollsters often ignore other indicators of opinion which may be much more revealing. The way a person lives may tell much about his or her attitudes.

The woman interviewed by Diane Bentley was unemployed. The man with whom she lived (the poll had categories only for "married" or "single") was a truck driver. There were four posters taped to her kitchen walls. One of them said more about her attitude to the energy crisis than her answers to any of the questions which Diane asked her. It showed a half-dozen greedy and bloated figures, each one representing a major oil company. Underneath, there was a caption: "Don't blame the truckers, it's these motherfuckers."

There were also signs of her habits. In addition to a refrigerator and a gas stove, the kitchen was equipped with a toaster-oven, an iron, a blender, and an AM/FM radio. In the next room there was a stereo and a portable television. For all her hostility toward the power industry, she was still its dependent customer.

This kind of background never finds its way into polls, and as a result, the full texture and shape of public opinion goes unreported. Pollsters behave as though opinion can be obtained only by asking questions, often artificially narrow ones, but in fact we express ourselves in many ways. Unfortunately, these clues to our attitudes and opinions are almost always lost on the pollsters.

Diane faithfully carried out her assignment. Though the woman's interest flagged after forty-five minutes, Diane kept firing off questions, carefully recording a response for each one. Not all interviewers are so diligent. The problem pollsters are least inclined to acknowledge is cheating by their interviewers, but it occurs fairly often.

The pollsters claim that they are very careful whom they hire and that they summarily fire anyone who fabricates data, yet the very nature of personal interviewing means that they have little control over their employees in the field. Most pollsters say that they routinely double-check a portion of all their interviews to be sure they were really conducted, but incidents involving both the Gallup Poll and the Harris Survey indicate that this validation process is not nearly so rigorous or effective as the pollsters would like us to believe.

In 1968 the *New York Times* commissioned Gallup to do an intensive survey of attitudes of Harlem residents. The information was collected, tabulated, and submitted to the *Times* for publication. An editor was so pleased with the poll that he decided to play it up by sending a reporter and a photographer to get a story about some of those who had supposedly been interviewed. At seven of the twenty-three addresses Gallup had given them, the newsmen could not even find a dwelling! Moreover, five other people who had allegedly been polled could not be traced—the addresses existed but apparently the people did not. Not even all the remaining interviews were legitimate. In one case the *Times* reporter learned that the interviewer had talked to four people playing cards and incorporated all their answers into one interview.

When the *Times* confronted Gallup with these false data, he ducked the responsibility for the phony interviews. He pointed out that the interviews in question had been submitted by two Columbia Law School students who had been specially hired for the project. There was no explanation, however, of why the Gallup Organization had not discovered the fakes itself.

Both before and after the incident, Gallup has repeatedly made the claim that his firm validates every interview made by new interviewers and every fourth or fifth interview done by one of their experienced employees. Obviously that was not done in this case. It was only accidentally that the *Times* uncovered the specious poll. Only sheer luck prevented them from printing it. Newspapers almost never double-check polls they print, but the Gallup incident suggests that perhaps they should.

The *Times* tracked down one of the special interviewers whom Gallup had tried to make a scapegoat. He claimed that he had actually talked to people "in the streets" but admitted that he had made up the names and addresses because he was under pressure from the Gallup Organization to complete his assignment—"I was uptight to get it in."

A year later Gallup's rival Lou Harris had exactly the same problem, but Harris was more fortunate in being able to keep his client from knowing it. Harris conducted a private survey in which several questions which had been contracted for were inadvertently left out. The client insisted on getting the information it had paid for, so Harris' firm sent out a special questionnaire by mail to those people who had supposedly been interviewed. According to a person who worked for Harris at the time, 25 percent of the questionnaires were returned as undeliverable: there were no such people or addresses. If Harris' firm had not erred by leaving several questions out of the original poll, the fabricated interviews would have routinely slipped through.

Even the most conscientious pollsters are limited in how much they can do to check the validity of interviews. They usually only call back people to see if they were interviewed. Checking the responses would simply be too difficult when one hundred or more questions have been asked, and, after all, the pollster may not really care if the responses are legitimate so long as they are plausible.

Albert Sindlinger is a self-styled maverick within the polling profession. He has all his interviewing done by telephone so that he can monitor his employees' work. He simply does not trust interviews done without supervision. Sindlinger says that he has occasionally hired Gallup and Harris interviewers to do his telephone canvassing, and from this experience he has a very low opinion of them. "They play the expert, but that's wrong. To do objective research, the interviewer has to be the dummy, not the other way around."

Sindlinger believes that there is a paradox which undermines personal interviewing. "If you hire somebody with any intelligence, after the tenth interview they're going to sit down and make up the other twenty-five. If you have somebody who is too dumb to do that, they're too dumb to record people's answers."

Given Gallup's experience with the Columbia Law School students in Harlem, Sindlinger may be right.

Another successful pollster, who asked not to be identified, says he knows several former Gallup and Harris interviewers who have admitted to him that they faked interviews regularly and were never caught. Like Sindlinger, he too now relies principally on telephone interviewing.

Gallup, more candid than most of his colleagues, admits that "every time you hire an hundred interviewers, you can be absolutely certain, whether it's in 1920 or 1990, that a small percentage of them will not do their work properly." The extent of fakery is open to question, however, and incidents like Gallup's Harlem poll suggest that the problem is more serious than most people believe. Oddly enough, the press does not seem very concerned. The *New York Times* continues to subscribe to Gallup's service.

There is another kind of fakery that pollsters seem oblivious to. Both Gallup and Harris maintain that their employees do not work so much for the pay as for the pleasure of getting out of the house and meeting people. Gallup says that his interviewers "don't earn very much." Harris' interviewers ostensibly start at $2.50 an hour, which seems very little for what can be hard work.

Unbeknownst to the pollsters, however, many of the interviewers pad their bills by inflating the number of hours they supposedly worked. Diane Bentley says, "I think of myself as an honest person, but it's just not realistic to do this kind of work for a couple of dollars an hour." Diane works fast and hard, then doubles or triples her actual hours when she sends a bill to her employer. On a typical day, she may earn seventy-five dollars; on her best outing, she earned a hundred and fifty dollars.

Diane is not unique. A former employee in Harris' supervisory office was sent out interviewing as part of her training. She submitted a bill for seventeen dollars for the ten interviews which she conducted; she was afraid that might have been a little high, given her inexperience. When she went to work in the office, she learned to her chagrin that the standard policy was to pay all bills on which the average cost of an interview was ten dollars or less. She could have turned in a bill for a hundred dollars and it would have been paid without question. "Maybe

Harris doesn't know what's going on, but everyone involved in the day-to-day operation does. People pad their hours all the time. Nobody cares. It's just a job."

Pollsters like Gallup, Harris, and Caddell differ in many respects, but they share an evangelical enthusiasm for their profession which may well blind them to the fact that many of their employees regard their work as just another nine-to-five job, an attitude which can obviously affect the quality of their surveys.

Sindlinger and others have responded to these problems by resorting to telephone interviewing, which allows them to monitor the work of their employees more closely. Most of the polling establishment, including Gallup and Harris, regards the use of the telephone as the sign of a minor league operation, claiming it has many built-in drawbacks which make it unreliable.

It is true, for example, that in the past the rate of refusal was higher for telephone interviewing than for that done in person. Recently, however, personal interviewers have encountered an increasing number of locked doors, so now the telephone produces an equivalent level of cooperation. Some pollsters also believe that it is easier to develop rapport with people in person and hence get more truthful answers from them, but this is hard to prove.

In large part the hostility to the telephone seems to be a legacy of the *Literary Digest* fiasco in 1936. The *Digest*'s sample was based in part on telephone books, which at the time made the sample heavily biased toward Republicans. That is no longer a real problem today. More than 90 percent of the households in the country have listed phone numbers. Those who do not fall into two quite different groups. First are the very poor who cannot afford a phone. Telephone surveys are thus inherently unrepresentative, but this is less of a problem than it might appear. Those who do not have telephones are also least likely to vote, so their absence should not materially affect political surveys. As a class they have negligible purchasing power, so they are of little interest to those conducting market surveys. It is also true that this group is often hard to contact by personal interview—because poll-takers tend to avoid poor neighborhoods—so there is not really the kind of avoidable bias which skewed the *Literary Digest* poll in 1936. The second group, those who have unlisted numbers, tends to be wealthy, but it is a small group, so its exclusion does not make a substantial difference. If necessary, they can be reached if the poll-taker randomly dials numbers.

Were it not for their desire to conduct long interviews, more pollsters would probably use the telephone. It is much easier to hang up a phone than to get someone out of your living room, particularly someone who has been trained in how to stay long after she has worn out her welcome.

Irwin Harrison, who polls for the *Boston Globe* and other papers, thinks that this is an illusory advantage. "You just don't need surveys that long. Whether it's a commercial client or a political client, they use only a fraction of the data you give them." A more limited study, he believes, can help the client understand what the real issues are, instead of being overwhelmed with page after page of computer print-out.

Harrison has found that in a fifteen-minute telephone interview, he can ask thirty-five or forty questions which will produce more than enough information for most clients. He is also justifiably suspicious of the information which personal interviewers elicit toward the end of interviews. "People get tired talking; all they're thinking about is how they can get the interviewer out of the house without being rude."

Pat Caddell strongly disagrees. Some of his interviews last as long as two and a half hours! He says the only real problem his interviewers have is getting people to stop talking. The practices of other pollsters, however, indicate that they realize that exhaustion does set in and that it affects the quality of the response. Gallup, whose surveys include questions for both his newspaper column and private clients, has said, "Our ballot always starts out with political and issue questions and then typically includes questions about products." This means that people are fresh when they answer questions for the Gallup Poll but less alert when they respond to the questions for the private clients, who, in essence, are getting the leftovers. Harris, by contrast, places his newspaper questions at various places in the interview.

There is no proof that one method of interviewing is superior to the other. It seems like a rather esoteric issue, but it is one which divides the polling profession. Gallup and Harris are contemptuous of those who use telephones, while Sindlinger and

Harrison think personal interviewing cannot be completely trusted. Each side is probably right in criticizing the other: both telephone and personal interviewing have significant weaknesses which can undermine the accuracy of a poll.

No matter how polling is conducted, in person, by telephone, or by mail, the pollsters have to take on faith that people are being honest with them. That faith may often be misplaced. A revealing statement recently appeared buried deep within an instruction circular the Harris firm prepared for its interviewers: "It's been brought to our attention that almost all of our surveys are showing the population to be more educated than what the census says it actually is. Nothing seems to be wrong with our samples, and there is no indication of error in recording or processing. Therefore, we feel respondents are exaggerating the amount of schooling they've had."

The language is blandly bureaucratic, but the fundamental meaning is nothing less than startling: the Harris organization admits that "in almost all" of its surveys, people have been lying!

Other studies have confirmed that, when asked, most people will say they are somewhat better paid and more highly educated than they really are. The pollsters have responded by making minor changes in their method of asking those demographic questions. For example, Harris interviewers now hand people a card on which different levels of education are placed in numbered categories. People are apparently somewhat more honest in assigning themselves to a category than they are in stating their education outright.

There is other evidence that people do not always tell the truth to poll-takers. In 1964 Elmo Roper discovered that a significant number of people would not say outright that they intended to vote for Barry Goldwater. When Roper told people to mark an ostensibly secret ballot, Goldwater invariably did four percentage points better than when they were asked directly for their preference. Goldwater ultimately did somewhat better than the pollsters, even those using secret ballots, had reported, which indicates that even more people were refusing to disclose their actual feelings to the pollsters.

In a number of recent surveys for various Democrats, Pat

Caddell has regularly asked people for whom they voted in 1972, Nixon or McGovern. In some states which Nixon actually won, a majority now say they voted for McGovern! Caddell says: "California was the first state in which we found this. McGovern won by an eight-point margin. We sent him a telegram saying, 'Demand a recount!' "

Caddell calls this an "anti-halo effect." The Watergate scandal made people want to disassociate themselves from Nixon. It was as if people across the country were buying bumper stickers which said, "Don't blame me, I'm from Massachusetts." Ordinarily, after an election there is a "halo effect" in which more people claim to have voted in favor of the winner than actually did. After John Kennedy was assassinated, it was almost impossible to find anyone who would admit that they voted against the martyred president in 1960, even though he had won by the barest of margins.

If people do not always tell pollsters the truth about how they voted in the past, then we must suspect what they say they are going to do in the future. In primary polls, George Wallace's strength has almost always been underestimated. People say either that they are undecided or that they are going to vote for someone else. According to Pat Caddell, the post-election studies which Yankelovich did in 1972 for the *New York Times* were unsuccessful in locating the full extent of Wallace's support. Even as people walked out of the voting booth, they still were reluctant to admit that they had voted for Wallace.

If people lie to pollsters about how much money they make and are not always truthful about whom they will—or did—vote for, they are also likely to be coy about giving their true feelings on controversial issues. In 1973, Charles Mee, a former editor of *Horizon* magazine, took an informal poll of several hundred people, asking them whether they favored impeachment of President Nixon. He discovered a higher percentage of people favoring impeachment than the Gallup and Harris polls indicated at the time. He concluded that people were afraid to stand up and be counted. Mee said, "The stunning and dismal finding of my little poll is that fear has eaten into the established, non-elective leadership of our country." People were afraid to express their true views, he believed, because they were genuinely worried about government reprisals.

Ordinarily it is difficult to measure the extent of lying that goes on in issue polling. Only the person who is answering the poll knows whether the opinion he expresses is truly his own. That people feel the need to be deceptive is of interest in and of itself, but unfortunately, from the bare statistics of a public opinion poll there is no way to tell who really believes what. The liars are lumped together with the truth-tellers.

In sum, though the theory of polling is scientifically sound, the actual practice is not. The practical problems facing the pollster are immense. More and more people are refusing to be interviewed and that makes opinion surveys less and less representative. In order to meet the requirements of the computer programmer, attitudes have to be forced into artificial and misleading categories.

Polling is not the pristinely pure scientific process the pollsters want us to believe it is. To know what really goes on as interviewers like Diane Bentley move from door to door is to demystify public opinion polling.

But as serious as these technical problems are, as suspicious of polls as they should make us, they are not the only source of polling error. The fundamental problem of polls lies in their analysis. Daniel Yankelovich is one of the few pollsters with the sensitivity and candor to recognize this. "I always get a little impatient when people suggest sampling is the problem. Sampling is the only scientific part of the field. It's not the source of the major errors. Yes, there are still problems, non-response for an example, but in tinkering with solutions, you are only going to improve the surveys marginally."

Yankelovich thinks it is in interpretation that polls become dangerously misleading. "That's where all the science ends. If you draw a perfect sample, but still ask dumb questions or analyze them erroneously, that's where the real problems are." Yankelovich suggests that improving the polls involves allocation of limited resources. "If you put your perfectionism and your resources into the sampling end, you have to be taking them away from the interpretive end, and that's where the gross errors are made today."

6

Influencing
the Public

George Gallup scoffs at the idea that published polls can affect
the outcome of elections. "I've never seen one shred of evidence
that polls affect voting behavior." Most pollsters agree. To say
anything else would be to admit that they are manipulating our
political system. But the fact is that published polls do have a
profound effect on voters. Indeed, polls can become the principal
issue in a campaign.

Nelson Rockefeller's late-starting bid to win the 1968 Republican presidential nomination is the most dramatic example of a
campaign which lived and died entirely according to the published polls. The 1968 campaign is an object lesson on how irresponsibly the press treats political polls. It illustrates both their
great influence and their unreliability. The year saw also the two
major polling organizations conspiring in a clumsy effort to
cover up their mistakes. The pattern of events which took place
in 1968 could easily occur again.

After months of vacillation, Rockefeller finally declared his
candidacy too late to enter the state primaries; hence he had to
persuade delegates who had already been selected that he was the
best choice. Many of the same delegates had booed him off the
convention rostrum in 1964; their natural loyalties were solidly
with Richard Nixon.

Rockefeller's only chance for the nomination was to prove that only he could beat the Democrats in November. Few Republicans wanted to repeat their experience of 1964, when they were on the wrong side of one of the biggest landslides in history. Richard Nixon, after all, had not been elected to office in his own right since 1950.

The odds seemed clearly against Rockefeller. The *New York Times* reported that "conversations with the delegates themselves suggest that they will abandon their inclination to give their votes to former Vice-President Richard Nixon only if the polls show conclusively that Mr. Rockefeller can win the election while Mr. Nixon cannot." If the results of the polls were the least bit ambiguous, the party regulars would resolve the doubts in favor of Nixon.

Thus Rockefeller's strategy was to prove that he was electable and Nixon was not. In a few short months he was to spend five million dollars on advertising—not to win votes in election booths, but to improve his standing in the nationally published polls. With the latest surveys in hand, he conferred privately with hundreds of delegates, reminding each one that the polls constituted the "hard facts of the situation today."

By June it seemed clear that the battle was going to be fought on Rockefeller's terms. James Reston declared: "The polls are likely in the end to be more decisive than the primaries and therefore the candidates will be campaigning to influence the polls from now until convention time in August. . . . Governor Rockefeller's strategy, if that is the word for it, is obviously aimed at George Gallup and Lou Harris." Even Nixon himself, who without real opposition had piled up a string of primary victories, admitted that the polls would be crucial. "The polls—that will be the drill down in Miami. Rocky will come in with figures showing he can run better than the Democrats in such-and-such a state. We of course will say that's not true. There will be polls flying all over the city."

By early summer Rockefeller's strategy appeared to be paying off. The Harris poll, based on interviews in the second week of July, showed Rockefeller ahead of Humphrey 37 percent to 34, while Nixon ran behind Humphrey by the same margin. The gap was small, to be sure, but it still suggested that Rockefeller was a winner and Nixon was not.

With the advertising campaign accelerating, the then New York governor's camp had good reason to hope that subsequent polls would show the margin ever widening, so that by August the Republicans convening in Miami Beach would have no choice but to heed the popular mandate. In mid-July, former Congress of Racial Equality director James Farmer released a survey which reported that black Republicans preferred Rockefeller over Nixon by 85 to 9 percent.

Rockefeller began to press his advantage hard. He sent telegrams to Nixon and to Republican National Chairman Ray Bliss asking for an impartial professional poll of each of the fifty states "to show where Mr. Nixon's strength lies." The party leadership turned down the proposal, which made it seem that they were afraid of what the results might show.

By late July there were signs that Rockefeller's strategy was reaching the delegates and that his novel campaign was on the verge of success. Less than two weeks before the balloting was to begin, the *New York Times* reported: "A number of delegates have begun to remark to interviewers that they are increasingly worried about Mr. Nixon's ability to win in November. They suggest that the latest Gallup and Harris polls, which have shown Mr. Rockefeller running more strongly than Mr. Nixon, have had an impact on their thinking." A county chairman from Washington state announced that he was switching from Nixon to Rockefeller "because I think he can win," and four others in his delegation who had earlier been pledged to Nixon said they now considered themselves independent. Erosion had begun, and Rockefeller's supporters were sure it would soon become an avalanche.

But suddenly the roof fell in. The latest Gallup Poll reported that Nixon had moved ahead of Humphrey, while Rockefeller could now manage only a tie! The governor's strategy of asking the delegates to abide by the polls had completely backfired. Herb Klein, Nixon's director of communications, said, "Mr. Rockefeller's polling game is all over." He was right. No matter what, Rockefeller could no longer conclusively prove that he was stronger than Nixon. The Nixon forces were jubilant, and Rockefeller gloomily admitted, "I don't blame them."

The impact of Gallup's single poll was tremendous. Had his

survey agreed with Harris', that is, if it had also showed Nixon losing to all the Democrats, with Rockefeller beating them, it is unlikely that Nixon would have won the nomination. Even as it was, the roll-call of states reached Wisconsin before Nixon was nominated; his projected delegate total was 672, only twenty-five more than the minimum needed for victory. In his chronicle of the 1968 campaign, Theodore H. White concluded, "One can go on down the ballot and easily find the slippages which might have reduced Richard M. Nixon's total by a hundred votes and thrown the convention into chaos."

If the Gallup Poll had depicted Nixon as a "loser," these slippages would have been all the more likely. If Nixon had been denied a first-ballot victory, Rockefeller might have picked up momentum, Ronald Reagan might have captured many Nixon votes from the South and West, or a dark horse might have moved in and grabbed the nomination. The fact that a single Gallup Poll could be so influential raises the obvious question of whether it was right.

The next Harris Survey, released only days after the Gallup Poll, reported completely opposite results. According to Harris, Rockefeller had now broadened his lead over Humphrey to 40 percent to 34 percent, while at the same time Humphrey had gained on Nixon, leading him 41 to 35!

These were exactly the kind of results the Rockefeller people had dreamed of, but they had come too late. Against Eugene McCarthy, Rockefeller held a 40 percent to 34 percent lead, according to Harris, while Nixon trailed McCarthy by a full eight points. Rockefeller's race for the nomination may have been over, but the disparity between the two best-known pollsters, if anything, brought even greater attention to the published polls.

Rockefeller made a vain attempt to salvage his campaign by pointing to the more recent Harris figures, but the unexplained discrepancies between the surveys meant that few delegates were going to be swayed by what any one poll said. The contest was no longer between Rockefeller and Nixon, but between Gallup and Harris. Which one of them was wrong?

Both polling organizations took a great deal of heat. The press demanded to know why we should ever trust polls when the two biggest pollsters come out with radically different results.

At the peak of the controversy, Harris apparently telephoned George Gallup, Jr.—the latter's father was in Switzerland—and persuaded him that they should make a joint statement for the good of the polling profession. The younger Gallup went along with the idea, and together they released a statement which argued that the differences between their surveys were not really as great as people seemed to think. They claimed that even though the polls were conducted only a week apart, "Public opinion changes over time, and each was an accurate reflection of opinion at the time it was taken." They also contended that normal sampling error could explain some of the apparent discrepancies.

To illustrate their second point, they presented a comparison of the surveys after they had been "adjusted" within the bounds of sampling error. By "applying a one-point shift" they proved the obvious, namely, that if you take away from Rockefeller's strength in Harris' polls and add to it in Gallup's, then the surveys will be somewhat closer in their results. Such a shift, they said, is "well within the bounds of normal sampling fluctuations." This juggling was sheer nonsense, of course, as it is equally possible that chance sampling error may have caused Harris to overestimate Nixon's strength and underrate Rockefeller's.

The press release concluded with the statement that the Harris and Gallup polls, together with unpublished surveys taken by Archibald Crossley for Rockefeller, showed that "Rockefeller has now moved to an open lead over both of his possible Democratic opponents, Humphrey and McCarthy."

This conclusion was in direct contradiction to the Gallup results, so many read it is a discreet confession by the younger Gallup that his figures were wrong and that Harris, whose later poll was more or less consistent with his first one, had been right all along. Although the phrase was carefully drafted, many people also construed it as implying that Harris and Gallup agreed that Rockefeller was stronger than Nixon, something the release did not say at all.

There is still some feeling in the polling fraternity that Harris fast-talked the junior Gallup into agreeing to the statement. George Gallup, Jr., has been active in the family firm for a num-

ber of years, but there is a sense among pollsters that although
he is a good guy, he has not acquired his father's savvy or tough-
ness. Many pollsters have been outspoken about the incident. At
the time, Burns Roper stated that there was "no Gallup data that
would support such a conclusion." Another pollster says of the
events leading up to the joint statement: "It was more a matter
of a basic weakness on George junior's part. He came out with
a set of results he didn't believe in, and Harris was confident
about his own."

In this confrontation between the two best-known names in
polling, Harris was cast as the heavy. He is still defensive about
the incident; his statements are inconsistent. He declared several
years ago that he had great regrets about the way he and young
Gallup handled the matter, but more recently he has said he has
no misgivings whatsoever. Harris now claims that Gallup initiated the idea of joint action and that Gallup was the one who
drafted the press release. That just does not seem plausible, however, in the light of how bad the statement made the Gallup Poll
look.

Instead of resolving differences, the statement generated still
more controversy. Journalists and politicians wondered aloud
about collusion among the pollsters. Herb Klein commented, "It
looks as if there's a pollsters' protective society being organized."
Congressman John Moss called for an investigation: "The time
has come for someone to step in and investigate the polling practices; the unprecedented act of the nation's two largest pollsters
combining in a common effort to correct their previous positions
underscores that need." *Time* solemnly concluded, "In persuad-
ing Gallup to endorse the apologia, Harris may have widened the
trade's credibility gap to the dimensions of 1948."

Even some pollsters agreed. Burns Roper was particularly distressed at the joint action by Harris and Gallup. "If this statement of an 'open lead' for Rockefeller is construed by readers as
being designed to influence the outcome of the Republican convention it will be most unfortunate, both for the political process
and for the public-opinion profession."

George Gallup, Sr., is sorry that the statement was ever issued
—"you can be damn sure nothing like that will ever happen
again"—but when he speaks of it he seems determined to protect

his son and to be fair to Harris. "It was a total misunderstanding on both sides; Harris was operating honestly and we were too. But it was one of those things which was well intended, but just looked bad when it came out in the papers." When pressed, however, Gallup bristles at Harris' claim that the statement was not his idea. "Harris called my son. You can be a hundred percent certain of that."

Burns Roper confirms the Gallup version of the incident. "The Gallup people were getting a lot of flak. The press was climbing all over them, asking for an explanation. When Harris proposed the statement, George junior was impressed with the need to close ranks and not to have a three-way fight with the press and with Harris."

Roper was very concerned at the time because "it looked like a whitewash, an attempt by the leading pollsters to cover up their discrepancies." Some in the Roper organization felt that they should take no public stand, and Roper himself admits he was worried about the "risks of a third pollster attacking his two competitors." Roper decided, however, that some pollster had to speak out, and he prepared his own release, which was highly critical of the Harris-Gallup statement. Out of courtesy, Roper called Gallup and told him of his planned announcement. Roper says Gallup was "very unhappy at that point, but he tacitly said to go ahead."

What is most striking about Harris' and Gallup's pre-convention polls in 1968 is that no matter how the discrepancies are explained, they raise serious questions about the pollsters' ability to report meaningful public opinion. Even if we accept totally the Harris-Gallup alibi that chance sampling error and the passage of time account for the dramatic differences, we are left asking what good polls are if they are subject to such wild fluctuations.

The discrepancies between the Gallup Poll and the Harris Survey figures persisted after the conventions. In late September, for example, Gallup reported that Nixon was leading Humphrey 44 percent to 29 percent, while the next Harris Survey had the margin only 39 to 31, a gap of only eight points as opposed to Gallup's fifteen. Throughout the campaign, Gallup made Nixon look much stronger than did Harris. Somebody clearly had to be off the mark.

Read carefully, the joint statement is no endorsement of the usefulness of polls. If the discrepancies which showed up so dramatically during 1968 can be casually brushed aside as the normal fluctuations of chance, time, and volatility common to all surveys, then we should never waste our time looking at polls. In essence, we are told that by the time the numbers are printed in the newspapers they may be hopelessly out of date.

There are two other explanations for the discrepancies. One is that the sudden spurt which Gallup reported in Nixon's strength was the hoped-for result of a special effort on Nixon's part to move the public opinion polls. His camp apparently had advance information on when the interviews for the final pre-convention Gallup Poll would take place.

Just before Gallup's interviewers went out, former President Eisenhower announced, on Nixon's signal, his endorsement of his former vice-president. Eisenhower had been silent about his preference up until then, and the announcement was doubly dramatic because Eisenhower made it while he was seriously ill and confined to a hospital. The endorsement was big news for several days and may well account for at least a few points of Nixon's temporary jump in support.

There is, of course, an even easier explanation for the difference between the polls, namely that some of them simply were wrong. Perhaps Gallup's sample was one in which the chance sampling error exceeded plus or minus three points. One out of twenty samples will be subject to greater-than-average error, and Gallup's may have been such a sample. Perhaps some interviewers got tired in the July heat and faked their reports. Wallace voters who did not want to reveal their true preferences may have been telling the pollsters Nixon one day and Humphrey the next.

There are a number of factors which could reasonably account for the differences, but it is pure speculation to say which was responsible. One error the pollsters made, however, is absolutely clear. They dug their own graves by lumping together those who said they would definitely vote for a particular candidate with the many other people who would only reluctantly express a preference one way or the other. Many Americans were dismayed by the possible choices, and millions of regular voters threatened not to vote. In forcing people to choose between Nixon and Humphrey or McCarthy and Rockefeller, the pollsters probably obscured the true state of public opinion, which was undecided and rather hostile to all the candidates.

There are many lessons which can be drawn from the published 1968 polls—how they influenced the Republican convention; how unreliable the polls appear when examined side by side; and how the two rivals, Gallup and Harris, clumsily tried to circle their wagons in defense against an understandably skeptical press. These lessons are significant, and they have relevance for us today.

But the most important lesson is also the most subtle and the easiest to overlook. In spite of all the commotion—the criticism from the press, the denunciations made by politicians, and the calls for investigations—in spite of all the uproar, nothing really changed. *Time,* which pontificated about a pollsters' credibility gap, still pumps tens of thousands of dollars a year into polling. Legislative attempts to regulate the pollsters have gotten nowhere.

The incident did not even make the pollsters more circumspect in their claims and pretensions. After his final pre-election survey came close to the actual results, George Gallup, Sr., had the chutzpah to complain, "We have had many words of praise even though there is a strange reluctance on the part of some media to give us full credit for our accuracy." This was barely three months after his firm had been ducking for cover after releasing its very questionable pre-convention poll!

If there was anything strange about the attitude of the media, however, it was how willing most of the press was to agree with Gallup's assessment that the polls had been right all along. In an editorial the *New York Times* congratulated the pollsters for their fine performance, concluding that "the accuracy with which they were able to forecast the outcome strengthens their contention they have managed to develop a reasonably exact science." The *Times* went on to declare, "The entire campaign provides a useful corrective to the argument that the polls exert an inevitably tyranical influence on the outcome."

In spite of being publicly embarrassed time and time again, the pollsters have no compunction about coming back and demanding that we pay them tribute. That the pollsters so easily get our congratulations when they happen to be right demonstrates how quickly we forget about the times they are wrong.

The pre-convention surveys of 1968 were not an isolated incident. Published polls have been central to campaigns before and since. Whether their influence can be called "tyrannical," to use the *Times'* adjective, depends perhaps on whose ox is being gored, but there can be no doubt about the magnitude of their impact.

Indeed, the syndicated surveys continued to have a great influence throughout the 1968 campaign. As mentioned earlier, Hubert Humphrey is convinced that adverse polls cost him dearly in contributions and workers; with a bit more support, at an earlier date, he thinks he would have won the election. The third candidate, George Wallace, also feels that he was victimized by the public opinion polls. Surprisingly, some pollsters agree.

The Republican strategy against Wallace was to hammer home the notion that a vote for him was "a wasted vote"; that is, by ignoring the real race between the two major-party candidates, Wallace supporters were letting Humphrey sneak in the back door. Wallace tried to counter this plea by arguing that even if he did not win a majority, he could force the election into the House of Representatives. Yet the advantages of this possibility seemed vague and confusing in the face of the clear-cut logic of the polls, all of which seemed to say that Wallace didn't have a chance of winning an outright victory.

Pollsters are usually reluctant to admit that their surveys can affect the way people vote, but Lou Harris himself believes that Wallace was the victim of polls in 1968. Wallace's ratings had grown from roughly 10 percent that spring to the teens by summer, and rose to the twenties by autumn, a good showing in a three-way race. But then, in the final month of the campaign, his support tumbled rapidly from 21 to 18 to 16 to 13 percent. Harris later described this drop as "a massive backlash against George Wallace, which I am frank to say was partly produced by our poll reports of his rising strength. . . . The voters took a hard front and center look at George Wallace—and recoiled."

Polls are neutral, we are told—except, of course, when they throw a knockout punch against someone of the likes of George

Wallace. Then the pollsters are happy to acknowledge their power.

Published polls were also the center of controversy during the 1972 presidential campaign. The influence of the national polls on Muskie's primary campaign in the spring and on McGovern's campaign in the fall have already been discussed at length. There can be little doubt that the high expectations set up by the early Gallup and Harris polls were the principal reason for Muskie's rapid downfall in the spring. The polls did not cost McGovern the election in November, but they certainly compounded his problems and distracted public attention from the substantive issues.

As more and more newspapers carry their own local public opinion surveys, the influence of published polls is extending beyond presidential elections to state and city campaigns as well. The *New York Daily News* poll sparks controversy in almost every election it covers. The 1970 New York senatorial campaign has already been discussed. Daniel Yankelovich reports that his private polls showed the bottom dropping out of Charles Goodell's campaign after the *Daily News* ranked him third in a three-way race.

Mario Procaccino says he has "no doubt" that he would have won the 1969 New York mayoralty race had it not been for the *Daily News* poll, and he may be right. The *News* reported that Lindsay held a twenty-one-point lead over Procaccino, yet Procaccino managed to finish just four points behind. Procaccino is sure that some of his supporters who just could not stomach voting for either of the other two candidates simply stayed home, while other followers voted for John Marchi, thinking that he had a better chance.

It is easy for a losing candidate to cry foul, but Procaccino had good reason for feeling victimized by the *News*. The *News* claims it has the world's largest circulation, and it usually prints its poll results on the front page in heavy capitals. The results of the *News* poll are picked up by other newspapers and television and given broad play.

For all its visibility, however, the *News* survey is quite crude by modern standards. Until quite recently, it was in essence a street-corner poll in which no effort was made to have a representative sample. The *News* did hire Lou Harris, as a consultant, not a pollster, to recommend changes in their techniques, but Harris is quick to disclaim the newspaper's work. During the 1970 senatorial campaign, Harris' results differed markedly from those of the *News*.

Although Procaccino made no specific accusations about the *News*, the experience left him "convinced that polls can be rigged and manipulated to convince unsuspecting voters in general, and the uncommitted voter in particular." Many other politicians, not all of them losing candidates, suspect that some newspaper polls are deliberately rigged to influence the outcome of elections.

House Majority Leader Tip O'Neill remembers several elections in which newspaper polls came up with results which conformed to their editorial policy but were out of line with O'Neill's finely tuned sense of public opinion. "A terrifically lopsided poll can kill a campaign by killing off the contributions." O'Neill strongly suspects that in some cases newspapers have rigged their surveys in order to stifle someone's candidacy, but without direct evidence to prove it he is reluctant to make specific accusations.

Proving that a given poll has been rigged is difficult because there are so many subtle ways to fake data. It is always possible, of course, for the pollster-in-charge simply to alter the final result, or use a biased sample, or ask loaded questions. Perhaps this is sometimes done, but only by amateurs, as the cheating would be obvious if somebody challenged the poll. A clever pollster can just as easily favor one candidate or the other by making less conspicuous adjustments, such as allocating the undecided voters as suits his needs, throwing out certain interviews on the grounds that they were with non-voters, or manipulating the sequence and context within which questions are asked. A number of tiny alterations could, when totaled, benefit the favored candidate by ten points or more.

It was just this sort of fudging that the *Denver Post* engaged in in 1948 when its own poll came up with the unbelievable news that Colorado was going Democratic. The editors were so intimidated by the national polls which showed Dewey ahead that they were afraid of standing behind their own research. According to Albert Sindlinger, Gallup made similar adjustments in 1948 whenever his poll results showed Truman in the lead, though Gallup vehemently denies the charge.

Polls can even be rigged without the pollster knowing it. If a candidate could get hold of a list of the sampling points, he could focus a high-powered campaign in those areas. If one were concerned only with influencing one's standings in the poll, then the rest of the country could be forgotten. Early in a campaign, when good ratings help to generate contributions, such a tactic would be very fruitful.

Most major polling organizations keep their sampling lists under lock and key, but usually a number of people have access to the information. A spokesman for the Harris organization has said that it would take the resources of the Russian secret police to be absolutely sure there is no wrongdoing. Paul Perry of the Gallup Poll has stated that there is not much the pollsters can do to prevent this kind of sabotage. "I think for anyone to get away with it on any large scale would be very difficult. Of course, you can do anything with money, I suppose. But anybody who's really an old hand at this business would begin to see something going wrong."

According to Frank Mankiewicz, who helped manage the McGovern campaign in 1972, his staff discovered many of the sample precincts which one of the networks was using for its Wisconsin primary polling. The McGovernites flooded these areas with workers and advertising in order to do better in the television poll. Mankiewicz says, "If you believe, as I do, that polls have a certain self-fulfilling component, then that action might have helped." Inflated poll results could be valuable not just for their hoped for-impact on voters, but, more important, for impressing potential contributers and the press. This kind of monkeying with the sample probably does not take place often, but when it does, unfortunately there is no way the careful reader can know of it.

Pollsters always adjust their raw data, for turnout, to weight an under-represented subgroup, and to allocate the leaners and the undecided. The thin line separating these sorts of adjustments from deliberate rigging is simply one of intent.

When people express concern about the misuse of published polls, their worry is usually rigged polls. Though rigged polls are a problem, the real danger lies with honest polls which are carelessly reported. We have already seen how crucial published polls can be to political campaigns. One would think this fact would make newspapers exceedingly careful in how they use surveys, but unfortunately a great many papers have become irresponsibly poll-happy.

The *Boston Globe* is a case in point. The *Globe* is a newspaper which takes itself seriously. In recent years, it has been ranked by some authorities as one of the ten best papers in the country. It has won more than its share of recent Pulitzer Prizes. The *Globe* is also a paper which cannot get enough public opinion polls. It subscribes to both Gallup and Harris, sponsors its own Massachusetts poll and New Hampshire poll, picks up other surveys from the wire services, and even runs the results of the postcard polls conducted by area congressmen.

Although the *Globe* has invested a lot of money in the local polls it sponsors, it does not have a particularly good record over the years. It was a *Globe* poll, conducted by John Becker in early 1972, which reported that Muskie had 65 percent of the vote in New Hampshire. Becker and the *Globe* can recite the old alibi that the poll was accurate for the time it was taken, but a truly accurate poll would have revealed that much of Muskie's apparent support was soft. Private polls conducted at the same time had detected that fact.

That same fall, a *Globe* poll of a local congressional race between incumbent Democrat Robert Drinan and Republican Martin Linsky sparked a great deal of controversy. A survey published on October 1 showed Linsky trailing by only ten points, well within striking distance, given some strong negative feelings about Drinan and the weakness of the national ticket. A second survey, however, published at the end of the month, showed Drinan with an overwhelming twenty-one-point lead. Linsky was understandably distraught; the last-minute poll could well persuade many of his supporters that theirs was a hopeless task.

Linsky was joined in his criticism of the poll by his opponent's campaign consultant, John Martilla, who said that his figures indicated that the race had actually remained stable during October. The Drinan people were afraid that if the *Globe* were to print another poll which reflected the true strength of the candidates, the results would be read as a slippage for their man, when in fact Drinan's support had remained constant.

The *Globe* editorial staff had private doubts about the accuracy of the poll even before they had printed it. They had checked with an independent pollster, who warned them not to publish it, but publish it they did. When both candidates attacked the legitimacy of the results, the *Globe* was compelled to take a third, unscheduled survey, and indeed it confirmed John Martilla's forecast: this one showed Drinan ahead by eleven points. Martilla's fears about how the poll would be reported were also correct. The *Globe*'s story stated that the "figures show that in those six days, Drinan lost 5 percent and Linsky gained 5 percent." There was no suggestion on the *Globe*'s part that its second poll might have been wrong.

After the voting, the *Globe*, in an article headlined "Poll Trends Borne Out in Bay State Election," congratulated itself on the performance of its polls, but a closer look at the figures shows that their surveys failed miserably. For example, Drinan, who had been given an eleven-point lead in the final survey, did win, but by a margin of fewer than five points. In the presidential race the *Globe* poll was pretty close in forecasting George McGovern's margin over Richard Nixon, but in the Senate contest it overstated Senator Edward Brooke's victory margin, reporting an incredible four-to-one lead over his opponent, when in fact Brooke won, but by less than two to one.

The *Globe*'s record in other 1972 congressional races was equally erratic, so much so that it appeared a dartboard could have produced equally good results. The poll was right in predicting a narrow victory for Gerry Studds, but it greatly overestimated John Moakley's victory over Louise Day Hicks, forecasting a nine-point gap, when in fact he won by less than two. The *Globe* poll had John Kerry, a Vietnam veteran who had spoken out forcefully against the war, ahead of his Republican opponent by ten points, but he ended up losing by nine!

That the *Globe* had the chutzpah to cite those results as evidence that its polls had been on the mark betrays its utter lack of responsibility. Though publicly the *Globe* management was expressing confidence in their pollster, privately they had misgivings; early in 1973 they let John Becker go and hired Decision Research to do their surveys. One of the principals in the newly retained firm was Irwin Harrison, who until mid-1972 had worked for Becker. Becker was not at all happy about losing the *Globe*, and there was talk that Harrison had stolen Becker's best-known client. Harrison denies the charge, pointing out that he had quit long before the *Globe* dropped Becker.

Many in the polling profession respect Harrison's ability more than they do Becker's, but Harrison has had his problems as well. In the fall of 1974 John Durkin, a relatively unknown Democrat, was running against New Hampshire Congressman Louis Wyman for the United States Senate. Pat Caddell, who was polling for Durkin, says his early studies showed that his client was suffering from a lack of visibility but that there was also a lot of latent hostility toward Wyman. By mid-October, Durkin had been able to generate publicity and exploit Wyman's weaknesses; at that point Caddell had Durkin running only eight or ten points behind, with strong potential to make up the difference.

The *Boston Globe*, through Decision Research, was covering the New Hampshire race. It, however, reported that Durkin was trailing by seventeen points. In the closing weeks of a campaign, a published poll which shows a gap of that magnitude can have a devastating effect. The Durkin people say that it totally dried up their contributions and had a stifling impact on their field-work.

Other politicians before John Durkin have blamed public opinion polls for their setbacks, but probably no candidate has ever had more justification than he. The election turned out to be the closest senatorial contest in history. Wyman won the preliminary count by several hundred votes. In a recount, Durkin was declared the winner by ten. On appeal, Wyman was again reported ahead, but this time by just two votes. Neither man was seated, and the matter was sent to the Senate. After protracted debate, it was unable to agree on a result.

Finally, nine months after the original balloting, Durkin agreed to a special election to resolve the question. Most political observers concluded that in so doing Durkin had graciously but

conclusively killed whatever chance he had of becoming a senator. Turnout in special elections is almost always low, which
usually benefits Republican candidates, as their supporters tend
to be more dedicated about voting. Nevertheless, Durkin won the
election and won it convincingly.

Durkin's ultimate success lends weight to his claim that the
Globe's poll cost him the original election. In spite of being
branded a sure loser, Durkin had after all managed a virtual tie
in the first election. That fact was not lost upon Democrats,
particularly labor groups, who gave him much more help the
second time around. Had it not been for the erroneous *Globe* poll,
however, the second election probably would not have been necessary.

The Durkin incident deserves a closer look to see why the poll
was wrong and how the newspaper treated it. When the survey
was first released, Durkin's pollster, Pat Caddell, was furious. In
looking at the supporting data, he saw that only one-sixth of
those included in the sample were independents, while actually
a third of New Hampshire voters are not affiliated with either
party. It was from independents that Durkin expected and was
getting heavy support. Durkin's supporters were thus under-
represented in the sample; in simple language, it was biased
against him.

Caddell is still outraged by the poll. "The *Globe* was irresponsible to print a survey for a state where a third of the voters are
independent and to only count them a sixth. I told the *Globe* this,
I told Harrison, too. The *Globe* runs around playing God on those
things. They got burned and they deserved to get burned."

Looking back now, Harrison seems to agree that his sample
was off; even before he released the poll he saw that if the independents were weighted according to their actual strength, the
results of the poll would be much closer. For that reason, when
he drafted the story to accompany the poll, he began with the
following sentence: "In New Hampshire turnout is the important factor in elections. If normal voting patterns prevail, this one
will be a toss-up." The editors of the *Globe* rewrote the story,
however, burying Harrison's cautionary words and giving prominent play to the supposed seventeen-point lead of Wyman.

Editors look for a flashy story and tend to play up the numbers and cut out the analysis. Here that tendency totally distorted Harrison's reading of the figures. Harrison himself does not criticize the *Globe* for its action, which is understandable, but the *Globe* badly used him in the process. Caddell thinks the blame must be fully shared by the newspaper and its pollster. "I don't care whose fault the story was. It was irresponsible for the *Globe* to print it and it was irresponsible for Harrison to let them do it."

Beyond the question of the technical quality of a published poll —whether the sample is biased, and so forth—is the matter of how the newspaper uses it. On this score also, the *Boston Globe* fails terribly, as indeed do many papers. The *Globe* has a disquieting habit of mingling its public opinion polls with its editorial policy. The *Globe* professes to draw a strict line between what is news and what is commentary. It keeps the syndicated columnists off the front page, whether they happen to be Mary McGrory or William F. Buckley. It does not, however, apply this policy to polls. It gives them front-page headline treatment, particularly, it seems, when the polls comport with the *Globe*'s editorial policies.

During the 1974 campaigns for the various state offices, the *Globe* gave heavy play to a series of surveys it took throughout the fall. Those candidates who were on the wrong end of lopsided reports angrily attacked the *Globe*. The *Globe*'s final polls proved to be the most controversial, because each candidate who was shown to be ahead also happened to be endorsed editorially by the paper. In a survey conducted on October 30 and 31, the *Globe* had shown Democrat Frank Bellotti with a comfortable ten-point lead over his Republican opponent, Josiah Spaulding, in the election for attorney general. But in a last-minute poll conducted on November 3 the *Globe* published figures which showed that Spaulding, whom the *Globe* endorsed, had taken a two-point lead. Bellotti eventually won the election, but by a narrow margin.

That the leaders in the final poll were all candidates endorsed by the *Globe* may have been just a coincidence. Irwin Harrison, their pollster, swears that this is the case, and he is an honest man. Although Belloti charged that the poll was rigged, he never came forth with any evidence of fakery. But even if the *Globe* was innocent of rigging, it was guilty of gross insensitivity about its roles. On one hand it was giving its support to certain candidates, while on the other it was promoting its supposedly objective surveys.

The *Globe* and other newspapers do not seem quite sure what public opinion polls are. They are not relegated to the op-ed page, or to the section where the other syndicated features appear. Instead polls are given front-page treatment, sometimes with eight-column headlines above the masthead. Yet they are not treated like other news, for the *Globe* will sit on surveys, saving them for the Sunday edition. Results for different races are released on succeeding days to draw out interest, instead immediately upon receipt from the pollster.

Like many other papers, the *Globe* is not at all discriminating about the polls it prints. It will give equal attention to a carefully constructed survey by some academic institution and a postcard poll by a local politician who has loaded the questions and used an unrepresentative sample. Most of the political columnists for the *Globe* are easy marks for politicians who want to leak favorable polls. Rarely do the columnists reveal who took the poll and under what circumstances it was conducted. A writer who is able to cite statistics, no matter what their source, apparently feels more sure of his observations.

A politician need only find a sympathetic columnist to whom to leak his poll and the results are sure to be played up. In early 1974, the *Globe* first published a survey sponsored by incumbent Governor Francis Sargent which showed him ahead of Democratic challenger Michael Dukakis by a seemingly unbeatable 44 to 24 percent. A little more than a month later, political editor Robert Healy devoted a column to a survey by an unidentified pollster which showed Dukakis ahead 45 to 37. Healy did not even mention the first poll, let alone try to reconcile it with the second.

A full-length column that David Farrell wrote in the summer of 1975 is an example of the extreme naïveté of some political writers. Farrell wrote that a recent survey conducted by Pat Caddell showed Edward Kennedy with an overwhelming lead over any possible challenger for his Senate seat. Farrell identified Caddell as a "nationally known pollster," but he did not reveal that Kennedy himself had been the sponsor of the survey.

So enormous was Kennedy's lead, Farrell wrote, that his campaign people did not want the press to get hold of the actual figures, for fear that potential Kennedy supporters would conclude that he was a shoo-in and thus would not contribute to his campaign. What Farrell did not say was that even he had been prevented from seeing the poll results. Farrell had taken it completely on faith that the poll showed Kennedy to be invincible!

Farrell refuses to say who his source was—he claims it was not Caddell—but whoever it was certainly exploited Farrell. Contrary to what the columnist had written, the poll had found areas of virulent anti-Kennedy sentiment. Kennedy himself privately confirmed that the news he received from Caddell was not all good.

In agreeing to ballyhoo the results of a poll without actually seeing them, Farrell had fallen for a gambit not unlike Richard Nixon's 1968 campaign promise that he had a "secret plan" to end the war in Vietnam. Supposedly we are living in an era of hard-nosed "show-me" journalism, but when it comes to public opinion polls, many political writers put aside their skepticism and swallow even the most outlandish claims hook, line, and sinker.

Not everyone at the *Globe* is happy with its infatuation with polls. Some discreetly complain that the money spent on sponsoring or subscribing to polls would be much better used if it went to hiring a couple of able reporters. Those in control of the paper, however, are firmly pro-poll, for both journalistic and personal reasons. Robert Healy, the political editor of the *Globe*, used to pester Becker's firm to get the latest results as soon as they were spewed out of the computer. This inside information would make Healy the center of attraction on the press bus. He would first drop little hints, then fragments of information, to his fellow reporters. Healy obviously loved the position of being in the know.

After all the controversy surrounding the 1974 election, the *Globe* announced that it was "re-evaluating" its policies and practices concerning political polling, but nothing really changed at that point. As an interested observer, Pat Caddell believes that no significant improvement will be seen in the paper's sponsored polls until it is willing to pump even more money into them. He suspects that budgetary limits explain why the *Globe* polls have run into trouble before; his answer is longer interviews, in person, instead of by phone.

Saying that the issue is really one of technical expertise, however, takes the newspaper off the hook. The *Globe* poll for the Boston mayoral primary in September 1975 was roundly criticized when incumbent Kevin White finished only eleven points ahead of his principal challenger, instead of winning by twenty-seven, as the survey had indicated. *Globe* editor Thomas Winship self-righteously pointed the finger of blame at his pollster, Irwin Harrison. "It is perfectly clear that the professional poll purchased by the *Globe* was wide of the mark on White's winning margin. Such a divergence calls for careful examination, and it is underway."

Poor Harrison was made the scapegoat even though polling primaries is a notoriously chancy game, especially in city elections where turnout is spotty and loyalties are weak. Harrison did take the poll, of course, but it was Winship who had put it on the front page. Harrison's warning that "White's wide lead may not necessarily hold up in Tuesday's preliminary" was buried eight paragraphs down in the story. One headline read, "Poll Shows White Holding a Commanding Lead Over Timilty." It thus was sheer hypocrisy for the *Globe* to place all the blame on their pollster.

Winship as editor of the *Globe* was also responsible for the paper's endorsement of Kevin White, so it should be Winship who is held answerable to charges that the paper once again manipulated its surveys to benefit its favored candidates.

Having been burned in the mayoral primary, the *Globe* finally altered its practices somewhat. In early 1976 it announced that it would not do head-to-head polling in the New Hampshire presidential primaries; instead it restricted itself to attitudinal polling, attempting to relate perceptions of the various candidates to people's stands on the issues. The shift was prompted less by a sense that it is wrong to try to predict elections weeks in advance than by a belief that primary polls are too crude to do the job. *Globe* editor Robert Healy says: "There is no way you can squeeze out those who don't vote, not accurately. When you're talking about a turnout of 70 or 80 percent in a final election, it doesn't make much of a difference. But when only 30 percent are turning out, it's a real killer."

The reoriented surveys were unusually imaginative. A month before the New Hampshire primary, the *Globe* reported that Reagan's workers had contacted approximately 81 percent of the state's Republicans and Republican-leaning independents, while Ford's staff had reached only 27 percent. The survey was an objective test of the activities of both organizations at that stage of the campaign. In Massachusetts, the *Globe* discovered that although Wallace had a substantial block of supporters, more than two-thirds of the Democrats and independents interviewed had unfavorable opinions of him, a radically higher percentage than that of any other candidate.

As enlightening as the *Globe*'s studies were, too many political writers still tried to reduce the figures to the trial heat format, describing those candidates ranked most favorable as being "in the lead." The *Globe* has also indicated that it may revert to head-to-head polling in future races; it continues to play up Gallup's and Harris' national polls.

The *Globe* is not unique. Scores of other papers are guilty of the same sins. If the *Globe* deserves special chastizing it is only because of its self-proclaimed aspirations toward responsible journalism. Other papers have been equally insensitive about their roles. In 1964, for example, the *Detroit News* used the same pollster as did Governor George Romney, who was seeking re-election. Romney's Democratic opponent cried foul, but the paper did not change pollsters.

In 1971 the *Philadelphia Daily News* devoted its entire front page to a poll for the forthcoming mayoralty race. According to the survey, Frank Rizzo had 30 percent; Hardy Williams, a state legislator, had 19; and Congressman William Green was a poor third, with 12 percent. The paper took pictures of the three candidates and blew them up proportionally to their supposed support, so that Rizzo's photograph appeared two and a half times as big as Green's.

According to Green, the impact of the poll was immediate and catastrophic. Before the survey, Green had felt he was gaining momentum; a fourth candidate had just dropped out, endorsing him. The weekend before the poll was released, Green received approximately forty thousand dollars in contributions. In the days after the poll appeared, he literally did not raise a dime.

What was astounding about this particular poll was it was

based on only 157 interviews! Philadelphia is the fourth largest city in the country, with a population of almost two million people. The sampling error for such a small sample may be plus or minus eight points. Thus it was entirely possible that Green, not Williams, was running second at the time and that he might have been within a few points of Rizzo.

Green insisted on a retraction. The *Daily News,* however, was convinced that it had taken an accurate poll. Green spent several days, first bringing in his own pollster, then getting nationally known pollsters to speak in his defense. Ultimately the editors were convinced of their error, and they magnanimously devoted an entire front page to an apology to Green.

Publicly Green says that he is satisfied with the way the paper treated him; it did go further than any other paper has under similar circumstances. People who have spoken to him privately, however, say that he is actually quite bitter about the incident, feeling that it permanently crippled his campaign. Apparently he does not express his true feelings out loud, knowing that in a feud between a politician and a newspaper, the newspaper always gets the last word.

Are published polls likely to improve, in terms of either technical quality or editorial treatment? Any improvement will require a basic change in editorial attitudes. Pollster Daniel Yankelovich criticizes what he calls horse-race journalism, which he thinks explains why most newspapers misuse polls. "Editors are looking for drama: who's ahead, who's behind. From their point of view that's the bottom line of politics." Yankelovich thinks that this orientation is peculiar to newspaper editors, and that the public at large really does not demand this kind of presentation.

The prime example of horse-race journalism is the great attention given to so-called trial heats, pairings of possible candidates long before an election. Too often the fact that one candidate is ahead in such a pairing is taken as an indication of his strength, but the figures are often misleading.

Peter Hart notes that in 1975 many columnists were concluding from the polls that, as weak as President Gerald Ford seemed in many respects, he still could beat any Democrat. Hart thinks tht kind of analysis, though all too frequent, is erroneous. "It's bad information. If Ford is beating Lloyd Bentsen by nine points, and only 31 percent of the people can really identify Bentsen, that's not good news for Ford." Hart says the press never looks behind the figures to see what they really mean. Few columnists, for example, ever consider the size or make-up of the undecided group, yet those voters often hold the key to an election.

Even in the closing days of a presidential election, when the Gallup and Harris polls get front-page coverage across the country, this horse-race journalism can be dangerously misleading. Assuming that the polls could be trusted to tell us how people are likely to vote, that information will not necessarily tell us much about who is going to win a presidential election. The Gallup and Harris surveys are national polls; they are not broken down state by state, so they are not a guide to how the campaign for electoral votes is going.

Recent presidential elections demonstrate clearly how little popular vote has to do with electoral strength. In 1960 Kennedy beat Nixon by a fraction of a percent in the popular vote, but dominated the electoral college 303 to 219. In 1968 Nixon and Humphrey were less than a percentage point apart in popular vote, but Nixon trounced Humphrey 56 percent to 35 percent in electoral votes. The difference in popular vote was virtually the same in 1968 and in 1960, but the electoral margin was much different.

Similarly, Johnson got a slightly higher percentage of the vote in beating Goldwater in 1964 than Nixon did in beating McGovern in 1972, but Goldwater managed to pick up three times the electoral votes McGovern did. Indeed McGovern, who received 38 percent of the popular vote, got far fewer electoral votes than did Wallace in 1968, even though Wallace received only 13 percent of the popular vote.

Thus, for all the attention they receive—and for all their influence—the national presidential polls tell us little about who is really ahead and by how much. Unless the electoral system is abolished, this will continue to be true. Yet the press continues to give greatest play to Gallup and Harris national polls, as misleading as they may be, simply because of the drama of two candidates battling for the meaningless distinction of winning the popular vote.

Yankelovich suggests that another basic problem with press coverage of the polls is that few journalists know how to interpret statistics. "They're terrible at it. It's not surprising, as it's not their field." Many reporters recognize their limitations in this regard and for that reason simply present the figures, ostensibly without interpreting them, but even this can be dangerous. Yankelovich says: "Where it's tricky is the situation when one candidate seems to be ahead by five points. You point that out, and you think you are just reporting, but you really are interpreting because you are attaching a significance to that figure and it may be very misleading."

Unless the press suddenly becomes more sophisticated, then, the only hope for more careful treatment of the polls lies with the pollsters themselves. Editors may have to defer to their pollsters to tell them what the meaning of the figures really is. A great many pollsters, including those already involved in syndicated work and others who do private polling, express dissatisfaction with the quality and orientation of existing newspaper polls, and some of them believe they could improve these polls. The optimists like Peter Hart believe they could move away from the horse-race, agree/disagree kind of question to get into deeper inquiries exploring the extent of people's knowledge and concerns, and the way that attitudes on different issues interrelate.

If past performance is a guide, however, that optimism is misplaced. The pressures are such that it may be impossible for pollsters to change the orientation of published polls. Lou Harris, for example, says it is unfair to criticize his surveys for being misleading when so much of what he releases to his subscribing newspapers gets edited out. "Some big papers now have a deliberate policy of boiling down all the poll stories to five paragraphs or less. Why I don't know. It emasculates the survey." Harris says he has better luck with the smaller papers—they tend to run his surveys in full—but he feels that the treatment of his polls ultimately rests with the whim of the editors. "None of them are going to give up their God-given right to say, 'I'm two inches short, so I'm going to cut this out.' There is no way you can command the newspapers to carry every last word you write."

Harris may be underestimating his power; with his reputation, he might be able to dictate the terms under which his polls would be printed. But for most other pollsters it is certainly true that the editors have final control.

Daniel Yankelovich has had a much better relationship with his client, the *New York Times*. Yankelovich says he has been allowed, even encouraged, to move away from simply "covering the horse race" and toward reporting on the dynamics of a campaign, such as how the issues and personalities interrelate. Although a typical Yankelovich report is not as punchy as a Gallup or Harris poll, it does describe the fabric of an election in much more detail.

There are still constraints, however, even if they are not editorial. The *Times* can afford to spend only so much money and devote only a certain amount of space to a given poll. Yankelovich admits that he often works on a dual standard. A private client can be given a book full of information and analysis, if necessary, but the *Times* can spare only a page at most.

The national newsmagazines have somewhat more money and space than most newspapers, and this is reflected on occasion by higher-quality surveys. In July 1975, for example, *Newsweek* published an extensive Gallup Poll on Edward Kennedy. Included in the tables was a rating of how well known other possible Democratic candidates were at the time. According to Gallup's survey, only 59 percent of the respondents had heard of Senator Henry "Scoop" Jackson, and other Democrats seeking the nomination were even less well known. Representative Morris Udall had a 38 percent rating, former Georgia Governor Jimmy Carter had a 26, and Texas senator Lloyd Bentsen lagged with 15 percent.

These ratings, it must be stressed, measured only name recognition. Had Gallup probed deeper to discover whether people could really identify the candidates and say something about them, the percentages would have been even lower. These measurements explain why Edward Kennedy had such a dominating position in the polls on possible Democratic nominees: in 1975 most Americans had not even heard of other possible candidates. These recognition ratings are essential if one is to understand the real meaning of the trial heats, but they never appear in the newspaper columns, including those by Gallup.

Of all the media, television, with its vast financial resources, can best afford intensive and careful polling, but it is also the medium which is least able to use polls as part of its news presentation. A typical story on Cronkite's or Chancellor's news program takes a few minutes at most, hardly enough time even to scratch the surface of a meaningful report. Moreover, television is preoccupied with news having visual appeal. It is hard to imagine anything which is less interesting to look at on a television screen than tables and graphs. As a consequence, the networks only report the horse race, who is ahead and by how much. Though they poll extensively, almost all of their research is used behind the scenes to determine the direction and shape of news coverage.

It is easy to knock the published pollsters for putting out misleading surveys; indeed the pollsters must be held responsible for what they do. But much of the blame must also go to the press, which has too often been negligent and insensitive in its own right. It is the editors who want the snappy headlines and who cut out the explanations of ambiguities and inconsistencies of opinion. A pollster who submits a survey which says that 30 percent are for candidate A, 20 percent for B, and the remaining 50 percent undecided may well be presenting the most accurate picture of public opinion, but he will not be in the newspaper business long.

The press has been hypocritical about public opinion polls. The polls get headlines, shape news coverage, and pump circulation, but when they go wrong, the newspapers are the loudest and most indignant in criticizing the pollsters for letting us down. Within a few months, when the dust settles, the newspapers are back again, promoting the polls as if nothing had happened.

In 1972 John Becker expressed some misgivings about the effect his early primary polls for the *Boston Globe* were having on the lesser-known candidates, but he saw nothing to be gained by stopping. "We're living in the age of the mass media, and if we and the *Globe* don't do it, somebody else is going to."

Becker is right. Polling is like any other business. So long as there is a demand for flashy, headline-grabbing polls, there will be a supply. Right now it remains a bullish market, with an apparently insatiable demand from the press.

7

The Polls
and Vietnam

Vietnam was the most distressing and divisive issue to confront Americans in a century. Feelings ran deep; events were perplexing and the future unsure. Reservoirs of patriotism were drained dry, and bonds which had served to secure our society were torn apart. The damage done may never be wholly repaired.

Though we watched the Watergate scandals unfold and were shocked that such abuses could take place in our country, we were in that instance largely witnesses to a drama played out by others. By contrast, we were all involved in Vietnam, if not as participants there, then as their families, their friends. As voters and taxpayers, we were inexorably drawn into the quagmire. In many respects, we were all its casualties. Watergate alienated Americans from their government, but Vietnam aligned them against one another.

Public opinion polls were at the forefront of the national debate about the war. For successive presidents, the war was as much a battle for support at home as it was a military confrontation in Vietnam. Support and opposition were painstakingly measured by the pollsters, who for more than a decade continuously monitored changing American attitudes.

Unfortunately, for all the attention the pollsters gave the war, they usually clouded the true state of public opinion rather than clarifying it. Vietnam provides a major case study in how polls can be manipulated and misinterpreted, and how dangerous the consequences can be.

Lyndon Johnson seemed to mark his success as commander-in-chief less by progress in the field than by the approval ratings in the various national polls. During meetings with diplomats or in the middle of speeches and press conferences, he would trot out the results of the latest surveys. "Mr. Gallup reported last week that we have gained 4 percent. Mr. Harris reports today that we have about 55 percent of the total in the country. Mr. Quayle made a nationwide survey and he shows about 55 percent."

Johnson's recapitulation of his standing at home sounded hauntingly like the Pentagon reports as to how much of South Vietnam had been "pacified." In the same way that more and more territory fell into enemy hands, Johnson saw his prized consensus slip through his fingers. When his ratings finally fell below the 50 percent mark, he stopped keeping score, at least publicly.

Johnson's preoccupation with the polls was legendary. He literally kept them in his back pocket, from which they could be pulled, as fast as Texas six-guns, to shoot down any criticism of his policies. Apparently his concern with polls went much deeper than just using them for propaganda purposes. In his recent book *The Anguish of Change*, Louis Harris recounts the impact of favorable polls on Johnson. Before the Tonkin Gulf incident in 1964, the Harris Survey had shown a majority dissatisfied with Johnson's handling of Vietnam, but after responding to the alleged attacks on United States ships, Johnson received support from 85 percent of the people Harris polled.

Johnson's aide Walter Jenkins called Harris to report that the president was greatly heartened by the results of the poll. Harris warned, as he had in his column, that such support could well be transitory and diminish in time. Jenkins responded: "Well, I'm sure you are absolutely right, but you know the boss. He won't forget this big show of support for a long time to come." According to Harris, Johnson regarded his showing in that particular poll as as much of a mandate as was the Gulf of Tonkin resolution passed by Congress!

British Prime Minister Harold Wilson told Johnson about deepening English opposition to American war policy. Johnson responded rhetorically, "Have you seen what the polls are saying?" Both publicly and privately, Lyndon Johnson regarded the polls as a national referendum, the results of which justified his actions.

When it was apparent even to him that the tide of public opinion had changed, Johnson acted like a jilted lover. By 1967 he was attacking the same polls he had embraced several years earlier, accusing the pollsters of giving aid and comfort to his enemies, both abroad and at home. "If some folks think a fellow is winged or crippled, they pile into him. If you've got an enemy, it gives them a lot of hope."

Throughout the war years, the government used polls not only to justify policy but to formulate it. Extensive intelligence operations in Vietnam included public opinion surveys of both ARVN and NLF defectors. The results were bad news, however, for the military agencies sponsoring the studies. The polls showed that, rather than winning the celebrated battle for the hearts and minds of the Vietnamese, we were losing it.

Our experiment in exporting democracy was an utter failure. The South Vietnamese in power discarded the best features of the democratic model—such as allowing legitimate opposition—and kept the worst aspects of modern campaigning. In 1971, for example, President Nguyen Van Thieu used intensive polling operations in his re-election efforts. The monthly surveys were conducted by Vietnamese but analyzed and paid for by Americans.

There were, of course, countless instances at home when polls were used for blatant propagandizing in the parallel battle to win American hearts and minds. People on all sides of the issue, right and left, exploited polls. A group called Peace Alert U.S.A., chaired by Senators Harold Hughes and Alan Cranston, among others, ran what they called a "National Peace Poll." The organization placed ads in various newspapers throughout the country asking people to respond to the question, "Should Congress bring the war to an end by cutting off the funds?"

At one point the group reported that "the vote" was running more than twenty-five to one in favor of cutting off funds. Such a result would not have been surprising, however, since those who would bother to respond to such an appeal would almost certainly be sympathetic to the sponsors' point of view. Had they really wanted an honest reading of public opinion on the question, they would have asked it of a representative sample of people.

This kind of propagandizing was by no means confined to the left. In 1970, a group which cleverly called itself the American Security Council reported that the great majority of "opinion leaders" in the country supported fighting for outright military victory in Vietnam. The group was headed by General Curtis LeMay and Clare Boothe Luce, among others. Questionnaires were sent to a limited group of people, the "opinion leaders" who had been identified as likely to be sympathetic to the organization's goals. Moreover, recipients were told to include ten dollars when they returned the questionnaire so that it would be "validated." The entire enterprise was aimed at producing pro-war results.

Partisans often used professionally-taken polls as ammunition for the debate on the war. In 1972, after President Nixon had ordered the mining of North Vietnamese ports, Senator Hugh Scott held a press conference to announce that a survey taken by Opinion Research Corporation of Princeton, New Jersey, showed that three out of every four Americans endorsed the president's actions.

Scott did not reveal, however, the actual wording of the question. The pollsters had asked, "To make sure that the supplies do not reach the North Vietnamese, the president has taken the following actions: (a) mined all North Vietnamese harbors; (b) had United States forces prevent any supplies from reaching North Vietnam by sea; (c) cut off all rail and other communications in North Vietnam to the extent possible; (d) continued air strikes against military targets in North Vietnam. Do you support the president's actions or don't you?"

The question was obviously biased in several ways. It presented the benefits of the action—cutting off supplies—without hinting at the possible risks, such as military confrontation with Russian vessels or the loss of planes and men. The final sentence would not have been much more loaded if it had read, "Are you a loyal American or a traitor?" Richard Nixon's name was carefully left out of the question.

Scott also failed to mention Nixon's close connection with Opinion Research Corporation. It was founded by Claude Robinson, who ran Nixon's polling operation during the 1960 presidential campaign. Robinson died shortly thereafter, but Opinion Research again polled for Nixon in 1968. It works almost exclusively for Republicans. Opinion Research refused to reveal who had sponsored the poll, saying only that it had been underwritten by "a private individual who wishes to remain anonymous."

Not only were the polls at the center of the controversy, the pollsters themselves were sometimes personally involved in political and diplomatic developments. In his book, Louis Harris reports being approached by a Russian diplomat attached to the United Nations delegation. Among the Russian's tasks was to report back on American public opinion, particularly as it was represented in the Gallup and Harris polls. In 1968, the diplomat contacted Harris with an outline of a peace proposal, ostensibly floated by his superiors in Moscow.

Harris carried the Russian message directly to Johnson, and later spoke at length with Secretary of State Dean Rusk and others in the administration. According to Harris, there was the possibility of a complete settlement of the conflict, but opposition by both the South Vietnamese and the Joint Chiefs of Staff killed chances of a total resolution. Harris claims, however, that the interchange was the foundation for the suspension of bombing in late October 1968 and the concurrent return to serious negotiations in Paris.

Direct American involvement in the war continued for another four years, and again Harris was involved in the course of negotiations. In 1972 peace talks were stalled, at least in part because the North Vietnamese and their allies were waiting for the outcome of the American presidential election; if McGovern were elected, they felt they could arrange much more satisfactory terms. In October of that year, Harris was contacted by the Russian diplomat, who asked that he meet with a high-ranking superior. The Russians grilled Harris at length about the presidential race. While Harris refused to make an outright prediction that Nixon would win, he did make it clear that Nixon's lead was substantial and that there was no sign of his losing much support.

At the same time Henry Kissinger was meeting with the North Vietnamese in Paris. Kissinger later told Harris that the North

Vietnamese had originally held to the same positions they had been demanding for months, but that when Kissinger pulled out the latest Harris Survey, showing Nixon with a twenty-seven-point lead over McGovern, the tenor of the discussion changed markedly. The North Vietnamese asked some questions about the poll, then adjourned. When they returned, according to Harris, they presented an entirely new proposal which became the basis of the final settlement.

Harris quotes Kissinger as telling him, "When they finally realized President Nixon would indeed be around another four years, they changed their tune and fast. Your poll carried a lot of weight." Harris and Kissinger both agreed that the North Vietnamese had probably checked back with the Russians in New York to confirm the authenticity of the poll.

It should not be surprising that polls were so central to the debate on Vietnam. To the extent that the war was a battle for American support at home, there had to be some test of public opinion. The military policies of both sides ultimately depended on an assessment of the American resolve to continue fighting, so polls became in a very real sense the measure of who was winning the war.

Polls were intensely scrutinized in part because all the other indicators of public opinion have such obvious limitations. A massive demonstration attended by two hundred thousand people still involves less than one-tenth of 1 percent of the country's population. Few cities had referenda on the war, and even where there was voting, mostly in college towns, the results were ambiguous. The wording on the ballot and the extent of the turnout could easily influence the outcome one way or the other.

Elections for public office are even more ambiguous. A peace candidate may defeat an ardent supporter of the war for reasons entirely apart from that issue. In the 1968 Democratic primary in New Hampshire, Eugene McCarthy apparently got the votes of many "hawks" who, impatient with what they regarded as Johnson's "no-win" policies, regarded McCarthy as an improvement over the existing situation, though certainly not a first choice.

Letters sent to congressmen likewise do not provide a fair test of public opinion. How opinion breaks down—how many are for and how many are against—depends very much on who receives the mail. When people write an elected official, they tend to write in order to confirm rather than to oppose his views.

For example, after the United States' invasion of Cambodia in 1970, Senator William Fulbright, an outspoken critic of the war, received thirty thousand telegrams. All but 750 of them expressed opposition to the action, a ratio of forty to one. Gerald Ford, then a Republican congressman and a staunch Nixon supporter, also reported that the preponderance of his mail was opposed to the intervention, but in his case the ratio was only four to one. Senator Barry Goldwater's mail, though very light, was even more favorable. This kind of pattern is common for congressional mail on all kinds of issues. Furthermore, counts of mail or telephone calls can easily be rigged.

Because other indicators could not clearly tell us how Americans felt about the war, we turned by default to the public opinion polls. That the issues were confusing and emotionally charged, and that it was not at all clear what we were fighting for, let alone whether we were succeeding, made the statistics all the more beguiling. In an atmosphere clouded with doubt and confusion, the crystal clarity of the polls was irresistible.

Yet for all the attention the polls received, they shed little light on the true nature of American public opinion on Vietnam. More often than not the polls compounded misunderstandings rather than resolving them. It may even be that the polls helped prolong the war by inhibiting political leaders from expressing their opposition in the face of what appeared to be broad popular support for American involvement.

The polls on Vietnam demonstrate both the enormous difficulty of surveying opinion on policy questions and the ways in which surveys can be manipulated and abused. A closer look at the Vietnam polls shows what a misleading picture of public opinion they often presented.

The pitfalls of election polling have been discussed in detail. For all the problems they present, however, election polls are a snap compared to polls on issues. In most elections, a person must either choose between two candidates or simply not vote. The alternatives are clear-cut. The voter may well have reservations about both candidates, but ultimately he must cast his ballot one way or the other. On questions of policy, however, there may be a vast range of alternatives to choose among, which makes the decision-making process all the more complex. It also means that when presented with questions on policy, people will be very sensitive to their wording, as small changes in phrasing can have broad implications.

Comparison of different surveys on Vietnam illustrates how critical the wording of a question is to the kinds of answers that will be given. In 1967, for example, two New York congressmen polled their respective constituents about the conduct of the war and came up with results that seemed radically different. The variation may have reflected actual differences in attitudes in the two districts, but much more important was the manner in which the polls were taken and the way the questions were asked.

Both surveys were done by postcard, and in each case only a small fraction of those who received questionnaires took the trouble to respond. Responses undoubtedly were affected by people's knowledge of who was asking the questions. One poll was conducted by Seymour Halpern, a Republican from Queens, the other by William Fitts Ryan, a Manhattan Democrat.

Halpern asked, "Do you approve of the recent decision to extend bombing raids in North Vietnam aimed at the strategic supply depots around Hanoi and Haiphong?" Sixty-five percent of those who responded said they supported the decision. At just about the same time, Ryan asked his constituents, "Do you believe the United States should bomb Hanoi/Haiphong?" By contrast, however, only 14 percent of his respondents supported bombing!

The New York newspapers gave a great deal of attention to the two polls, in part because of the apparent conflict between them. Some people read the two surveys as confirming the theory that opposition to the war was coming essentially from upper-middle-class liberals, such as those in Ryan's district, while middle-class Americans, such as those living in Queens, were solidly behind the administration.

In truth, however, the markedly different responses simply reflected differences in the wording of the questions. In essence, people in Queens had been asked something quite different from people in Manhattan. Halpern asked if bombing should be "extended" to include "strategic supply depots." The implication was that this was just a modest expansion of existing policies, and a defensive one at that, not aimed at the cities themselves. Equally important, Halpern's question asked about approval of a "recent decision"; in effect, people were asked to ratify an action which the government had already seen fit to take. By contrast, Ryan's question was hypothetical. For people who were unaware of day-to-day developments in the war, the question could be read as suggesting a possible future course of action. Some pollsters have noted that when questions are posed in the abstract, people have a natural tendency to say no. Ryan's question hit this bedrock negative response.

That the responses to questions on the same basic issue can vary so much according to wording obviously suggests great possibilities for manipulating poll results. Whether Ryan and Halpern deliberately loaded their questions or were simply oblivious to nuances in wording is not clear. Most congressmen know very well how to phrase a question to get the responses they want. But the intent of Ryan and Halpern really does not matter, for in either case, the end result of their polls was to confuse public opinion rather than to clarify it.

We should not be surprised, perhaps, that partisans would manipulate questions in order to produce results they can exploit. What is astounding, however, is the carelessness demonstrated by many professional pollsters in their drafting of polls on issues. For example, in June 1969 the Gallup Poll asked, "President Nixon has ordered the withdrawal of twenty-five thousand troops from Vietnam in the next three months. How do you feel about this—do you think the troops should be withdrawn at a faster or a slower rate?" Faster withdrawal was preferred by a plurality, 42 percent, while 16 percent said they favored slower withdrawal. Twenty-nine percent, however, refused to accept either alternative offered, and instead volunteered that they supported the president's announced withdrawal rate.

Within several months, the Harris Survey asked the same general question; however, it specifically offered not two alternatives but three: faster, slower, and about right. Faster was again preferred to slower, 29 percent to 6 percent, but this time 49 percent said the president's withdrawal rate was "about right."

People who read both polls might well have been left with the impression that Nixon's policies had gained markedly in support, moving from second in Gallup's survey, with 29 percent, to first in Harris', with 49 percent. In fact, however, the differing results were likely less a result of the passage of time than they were the product of a slightly different wording. Whoever had drafted Gallup's question had neglected to consider that people might well support the status quo. By presenting only two alternatives, faster or slower, the question made it more likely that people would appear to disagree with Nixon's rate of withdrawal. Those who supported the status quo had to volunteer that information in the Gallup Poll, while the Harris Survey explicitly presented that possibility.

In the spring of 1973 both the Gallup Poll and the Harris Survey reported that a broad segment of the American public was opposed to bombing in Indochina, but the surveys showed quite different measures of the extent of opposition. Gallup had asked: "As you know, U.S. planes are bombing Communist positions in Cambodia and Laos. Do you approve or disapprove of this action?" Twenty-nine percent said they approved, but 57 percent said they did not.

Harris asked: "The U.S. has used B-52 bombers in Cambodia in the fighting that has gone on there, because it is felt that the peace in Vietnam is threatened. Do you approve or disapprove of the bombings by U.S. planes in Cambodia?" Thirty-three percent favored the bombing, while 49 percent disapproved of it.

Thus, according to Gallup, a two-to-one majority was opposed to the bombing, while Harris reported only a three-to-two plurality. By stating a rationale for the bombing, Harris' question apparently was able to elicit more support for the government's policy.

Both Gallup's and Harris' questions were flawed in a subtle but important way. Apparently the questions were asked of everyone in the sample, whether or not they had heard or thought about the United States attack. Responses of people who paid no attention to the matter until the pollster raised the question were lumped together with those who had already given the issue careful study. An honest picture of public opinion would show the number of people who had never considered the question. To seek a response from everyone creates the erroneous impression that more people are concerned about an issue than is really the case.

Even with an issue as crucial as the war in Vietnam there was surprising ignorance and apathy. By March 1966 there were hundreds of thousands of American fighting men in Vietnam, and the war had dominated the news for two years, yet a poll taken at that time revealed that only 47 percent of the American people could name Saigon as the capital of South Vietnam, and even fewer, 41 percent, could name Hanoi as the capital of the North. In response to another poll in 1967, just 48 percent of the public said that they had a "clear idea of what the Vietnam war is all about." The latter response, of course, may be as much a reflection of the confusing nature of the war itself as a measure of public ignorance.

In order to get significant readings of public opinion, a pollster should weed out those people who have not given any thought to an issue, in much the same way that they weed out from their election polls people who are not likely to vote. This can be done through a series of "filter questions" which test the extent of the respondent's knowledge of an issue before asking his or her opinion on it.

Variation in the wording of questions, and the corresponding variation in results, was the rule, not the exception, during the entire course of the war. In April 1975 Gallup reported that Americans favored giving the president authority to use United States troops to evacuate American citizens from South Vietnam by an overwhelming 76 to 20 percent margin. The very next day, however, Lou Harris reported, "A 68 to 22 percent majority of Americans oppose sending any American troops into South Vietnam to help evacuate U.S. citizens or Vietnamese allies whose lives may be endangered by a Communist takeover."

The results seem entirely contradictory, but are principally the product of nuances in the way the questions were phrased. People who saw both polls may at least have realized the complexities of opinion on the subject, even if the polls were not very helpful in defining its shape and substance. Those who saw only one poll, however, were left with the erroneous impression that

Americans were either wholeheartedly in favor of using troops for evacuation or adamantly opposed to it, depending on which poll they happened to read.

This kind of inconsistency in results is typical of all issue polls, particularly those on complicated problems like Vietnam. It is a tremendous obstacle for anyone who is seeking to find consistent threads of opinion from which the whole fabric of American attitudes on the war can be woven. But if it is a burden for the serious historian, it is a boon for the pollsters, for they can always find an alibi to justify their results: either opinion has changed over time, or, if contradictory polls appear at the same time, the differences can be explained away as being caused by differences in the wording.

It is thus impossible for a pollster to be proven wrong on an issue poll. Gallup can say that Americans support evacuation by almost five to one, and twenty-four hours later Harris can say that they are opposed three to one. If there were a comparable disparity in the case of an election, one of the pollsters—perhaps both—would be proven wrong. On a policy question, however, there is no other test of opinion beyond the poll which can prove embarrassing. Little wonder, then, that most pollsters say that they prefer polling on issues to doing election surveys.

Because it is next to impossible to correlate the results of the countless different polls on Vietnam, most analysts who have tried to trace the development of American attitudes on the war have done so by looking at one isolated question and seeing how responses to it varied through the years. Unfortunately, this kind of trend analysis often produces superficial and questionable conclusions.

For example, the *Public Opinion Quarterly*, the journal of the American Association for Public Opinion Research, published a lengthy article in the spring of 1970, the thesis of which was that "opposition to the Vietnam war is more widespread in the older generation than among youths." To establish this somewhat surprising proposition, the author cited responses over time to one particular Gallup question, "In view of the developments since we entered the fighting in Vietnam, do you think the U.S. made a mistake sending troops to fight in Vietnam?"

The results showed that from early 1966, when the question

was first asked, through 1969, the age group fifty years and older always had the highest percentage of those who said entering the war had proven a mistake. Their percentage was consistently higher than the figure for the group thirty to forty-nine, as well as that for people in their twenties. The fifteen-page article contained a great number of tables in which the results were broken down according to sex, income, profession, and education, and comparisons were also made with similar polls about United States involvement in World Wars I and II and Korea.

Yet for all the care that went into the preparation of these exhaustive cross-tabulations, one fundamental fact was overlooked. The author simply assumed that anyone who said entering the war had been a mistake must also be opposed to it. People who regret getting married do not necessarily get a divorce. Likewise, those who think it was a mistake to get into the war might consistently believe that once the United States was involved, it had to see the war through to the end. Gallup's question did not really measure how many people were opposed to the war; therefore it was erroneous to conclude that opposition was greater among older people.

A 1968 Survey Research Center Study showed clearly that regarding entry into the war as a mistake was not equivalent to being opposed to it. According to the poll, a five-to-three majority considered the original commitment a mistake, a result which is consistent with the Gallup finding for the same period. Among those who thought entering the war had been an error, there was a slight tendency to lean toward de-escalation as the best policy for the future, but this tendency was surprisingly weak. Almost as many of these people favored escalation! Specifically, although a majority of the sample had called the war a mistake, a majority almost as large had endorsed taking "a stronger stand even if it means invading North Vietnam" rather than supporting complete withdrawal.

This error of reading far too much into the responses to one specific question was repeated time and time again. In their book *The Real Majority*, Scammon and Wattenberg made this mistake and compounded it when they concluded, "Young people were *more* hawkish than the over-fifty generation." As evidence for this claim, they cited another Gallup Poll, this one taken shortly

before the 1968 presidential election, in which people were asked
to identify themselves as either "hawks" or "doves." Gallup had
reported the following figures:

	Under 30	50 and over
Hawk	45%	40%
Dove	43%	42%
No Opinion	12%	18%

Leaving aside for the moment whether the Gallup question
can be considered a meaningful test of attitudes on the war,
Scammon and Wattenberg's flat-footed conclusion simply is not
supported by the numbers. The responses of those under thirty
were virtually identical to those of people over fifty. Both were
just a point away from being an absolutely even breakdown. The
difference of a few points could well have been the product of
chance sampling error. The data as it appeared, however, fit the
Scammon and Wattenberg thesis, and they gave no warnings that
it was not conclusive.

Misinterpretation of poll results was also frequent because few
people realized how sensitive opinion is to acts of leadership and
changing events. Percentages which seem clear-cut and decisive
may actually reflect only vaguely defined and loosely held attitudes. Forceful policy decisions can have a dramatic effect on poll
results—what seems to be minority opinion one day can be gal-
vanized into majority opinion the next. Unfortunately, those
who look to polls for mandates often forget this.

In March 1968 the Gallup Poll asked, "The North Vietnamese
have said that if we agree to stop the bombing, they will agree
to the peace negotiations. How do you feel—should we stop the
bombing or not?" Fifty-one percent said they opposed stopping
the bombing, while 40 percent favored a halt.

Several weeks later, however, President Johnson announced
that he was suspending the bombing of more than 90 percent of
the territory in North Vietnam. Immediately, Gallup surveyed
the public's reaction to his decison. Sixty-four percent now said
they approved of the suspension, and only 26 percent said they
did not. What had been a minority had become, virtually overnight, a five-to-two majority!

This dramatic change in the polls did not necessarily mean that scores of millions of Americans had given careful thought to their earlier positions and then had revised their basic points of view. Before Johnson's announcement, the issue was hypothetical and public opinion was correspondingly vague and shallow. His decision and the dramatic circumstances it precipitated changed all that. The idea of a bombing halt gained legitimacy by being translated from abstract proposal to concrete action. The two polls, read together, show that even at this late date many Americans were willing to defer to the administration's judgment. If the government had not ordered a halt many people would have been opposed to one, but if the government saw fit to stop the bombing the same people would concur.

Thus the March poll, ostensibly showing a majority against a halt, was not a popular mandate to continue the attacks, and Johnson's decision to stop them was not a case of a politician flouting public opinion. Opinion responds to acts of leadership. Yet politicians who had earlier argued for de-escalation had to confront polls, like the one Gallup took in March, which seemed to say that such efforts were opposed by a majority of Americans. Polls, in essence, are the ally of the status quo.

On countless other issues, polls have recorded public opinion as being opposed to one suggested innovation or another, when in fact people really mean that they are willing to go along with whatever the government deems wisest. Once the innovation is undertaken, it is immediately supported. This pattern was particularly true for Vietnam policy. After analyzing many surveys on Vietnam, Philip Converse and Howard Schuman of the Survey Research Center at the University of Michigan wrote that "support of the president seems to rise after any new initiative, whether it is in the direction of escalation or a reduction of commitment."

That acts of belligerence and acts of accomodation would both produce spurts of approval may have reflected a general sense of "my country right or wrong." As time wore on, it became part of a repeating cycle of hope and frustration. As it became increasingly obvious that the end of the war was not near, people would grow restless with existing policy. Any possibility of breaking the stalemate, whether through new negotiation efforts or heightened attacks, was welcome. When in time it became clear that the new initiative had been no more productive than the ones which had preceded it, Americans would again become impatient and be ready for another shift in the course of the war. Hopes can be dashed only so many times before cynicism sets in. The general drift in support was downward, yet throughout the war there would continue to be spurts of approval whenever a president took affirmative action, no matter what it was.

There is still another factor which went into the dramatic shift in Gallup's 1968 bombing polls. Much of the change may be ascribed to public sympathy for Johnson. Approval of presidents almost invariably goes up in times of stress. Eisenhower's approval rate went up during the Suez crisis in 1956, but it shot up even higher after he was publicly embarrassed by the Russian capture of an American spy plane on the eve of a scheduled summit conference in the summer of 1960.

Kennedy's popularity was at its highest not after negotiating the removal of Russian offensive missiles from Cuba in 1962, but after the disastrous Bay of Pigs invasion the year before! On seeing the poll, Kennedy wryly said: "It's just like Eisenhower. The worse I do, the more popular I get." So it probably was with Johnson. Some of the support for his actions may have been a carryover from pity, perhaps guilt, which people felt toward him after he announced that he would not run for re-election.

The sympathy or loyalty factor which won support for Lyndon Johnson is always reflected in the polls which test general approval of the president, but this sort of approval does not always extend into support for his specific policies. This can sometimes create an anomalous situation in which the public expresses its approval of the president's general conduct of his office on one hand, but on the other rejects the very things which he is doing.

Public reaction to the Cambodian invasion in 1970 is a dramatic illustration of this phenomenon. Immediately after President Nixon announced his invasion plan, the Gallup Poll took a special survey to test public opinion. Gallup reported that 51 percent said that they "approved of President Nixon's decision" in Cambodia.

The administration played the result for all it was worth. The silent majority had spoken, and it was backing the president. Members of the press read the poll the same way. Yet in the same survey Gallup's interviewers asked people whether they approved of sending American troops into Cambodia. This time, when the specific action was identified but the president's role was not mentioned, the results were entirely different. Fifty-eight percent said they disapproved of United States military intervention in Cambodia, while only 28 percent supported it!

When specifically asked about presidential action, a slim majority had still been willing to pay lip service to the chief executive's prerogatives. But when asked about the specific policy which had been announced, they could not bring themselves to endorse it. That Nixon could muster only a tepid show of support was a clear sign of how war-weary the American people had become. A year later, Nixon's approval actually went down, from 56 to 51 percent in the Gallup Poll, after he ordered the invasion of Laos. His action was so inconsistent with his announced policy of disengagement that Americans could no longer muster even a token increase in support.

Lou Harris thinks that one of the legacies of Vietnam will be a lasting cynicism among Americans. In his view, we will no longer be willing to grant our presidents a blank check of approval. "What happened with Vietnam, among the casualties, was the crumbling of our old belief that, other things being equal, we would always give the president the benefit of the doubt in times of crisis." From now on, according to Harris, people are going to say: "I'm from Missouri. The president is going to have to prove his case to me."

This notion that Vietnam, and Watergate as well, has spawned an age of cynicism and distrust of government has become the conventional wisdom. The experience in Vietnam certainly shows that the well of public faith in the government is not bottomless. Yet there are already signs that we are putting that skepticism behind us. In the spring of 1975, for example, the Harris Survey reported that approval of President Ford had shot up from 40 percent positive, 57 percent negative, to 50 percent positive, 49 percent negative, after he sent Marines to rescue the cargo ship *Mayaguez*, which had been seized by the Cambodians. In spite of reports that the Cambodians had expressed a willingness to settle the matter diplomatically before the United States operation was undertaken, and even though more men died in the rescue attempt than were on the vessel, Harris reported that 79 percent of the American people approved of Ford's handling of the affair.

Even though these temporary spurts of approval occurred time and time again during the Vietnam war, we still have not learned how transitory this kind of support is. Political pundits on both the right and the left concluded that Ford's decisive action on the *Mayaguez* had greatly strengthened his chances for election in 1976. Ford's increased popularity in the political polls supposedly showed that the Democrats would have a hard time finding a candidate to beat him.

Two months later, however, after all the news of the *Mayaguez* had been forgotten, Harris again reported on Ford's popularity. It had dropped back down to 41 percent positive, 56 percent negative, almost the same level it had been at before the *Mayaguez* incident. In the end, the event apparently did little to bolster Ford's basic support.

The polls on attitudes toward Vietnam were often misleading or misinterpreted. One would think that the mass of conflicting results would have caused the polls to neutralize one another. Nevertheless, they had a profound effect on the way we thought about public opinion, and as a consequence, they had a correspondingly large influence on the development of American policy.

That the polls seemed so contradictory—the public would apparently be opposed to a bombing halt one week, yet in favor of it the next—made opinion watchers look for some simpler key to what Americans really thought. For politicians and reporters alike the question was whether people supported the president's policies. That was the bottom line. Was the president winning the battle for support at home or was he losing it? The question of whether people had confidence in the president's handling of the war became a national referendum on Vietnam policy.

Daniel Yankelovich thinks the focus on this question obscured the depth of opposition to the war. In the early polls, he notes, there were already signs that Americans were doubtful about administration policy and favored a negotiated settlement. Yankelovich says: "If you read the fine print, the detailed data, you saw all kinds of signs of hesitancy and doubt. But that's not what the headline writers picked up. They played up the question of presidential support."

Yankelovich believes the question "Do you have confidence in the president's conduct of the war?" was a very poor indicator of American attitudes. "If you ask an employer whether he has confidence in such-and-such an employee, he is going to answer yes until he is ready to throw him the hell out. He may have all sorts of reservations. But confidence? That's the last step."

The question, then, that was regarded as the most important test of American opinion on the war was in fact one which would elicit support for the administration as long as possible. A vote of no confidence would necessarily be the last straw. Yet our preoccupation with this single measure would for years divert our attention from other signs that the administration was out of step with the people.

These confidence polls, which long seemed to say that people supported administration policies, inhibited those in Congress who had their doubts about the war from speaking out against it. Representative Peter Rodino says: "To say that the polls didn't have an impact would be not to face reality, not to understand that political animals look everywhere to form a judgment. Unquestionably, some were influenced by public opinion polls, particularly from their own district."

If the polls had been telling congressmen what people really thought, that would be one thing. But if, as Yankelovich suggests, there was a lag between what the people were feeling and what the most publicized polls seemed to say, then the political animals were on the wrong trail. In their efforts to follow public opinion, they were actually going against it.

It was not just supporters of the war who gave the polls credence. George McGovern, who was critical of United States policy early on, says: "I followed the polls very closely on the war, and I think they tended to be right. They showed high approval of administration policies until the late sixties, when it began to drop." Yet at the same time as McGovern acknowledges the general-confidence polls as the basic test of American public opinion, he also recalls that other polls tended to cut the other way. "I remember a Gallup Poll which showed that people supported the McGovern-Hatfield amendment, calling for withdrawal of American troops within ninety days. But we lost the vote on the Senate floor, so that may have been an instance where the Senate was lagging behind the people."

House Majority Leader Tip O'Neill strongly disagrees with the idea that the American people were opposed to the war before it was obvious in the polls. By his own description, O'Neill was an "avid hawk" up to 1967. After being briefed by teams from the Pentagon and the State Department, he said he would go anywhere to speak in defense of administration war policy. Only after being asked by a student if he had been briefed by the other side did he reconsider his views. He spoke with many second-level bureaucrats and junior officers who publicly were endorsing the war effort, but who privately told O'Neill that the situation was already hopeless, and, if anything, was getting worse.

O'Neill sent 150,000 newsletters to his constituents telling them that he had changed his mind on Vietnam. The antagonism his action prompted convinces him that most people were still in favor of the war. "I had people who had been friends of mine for years turn their back on me or cross the street to avoid me. I had to go to every communion breakfast and Rotary meeting to turn the sentiment around."

O'Neill says that he took a lot of heat but that eventually those he had alienated came back into the fold. He gently mimics "old Irish ladies" who have come up to him and said: "God bless ya, Tip. You know I was so mad when you turned against Johnson that time, Tip. I says, what's he done? After all these years our faithful friend and now he's runnin' around with that Harvard Square crowd. Oh God, Tip, I said a prayer for myself for thinking harm about ya. How proud I am that you were right so far ahead of the rest of them."

The reaction O'Neill encountered was largely a product of his own doing. He had, after all, spent years speaking out loudly and often in favor of administration policies. Many of his constituents had listened and deferred to his judgment. Small wonder that when he reversed his position there was some resistance, particularly at the outset. The public opinion polls had continued to show apparent support for the war because respected leaders like O'Neill had been in favor of it. Too often, however, politicians bowed to these polls, and an incestuous circle was created.

O'Neill may well have taken heat when he announced his shift, but the nerve he touched was not so much support for the war as it was hostility between town and gown. The war was a polarizing experience, and nowhere were the lines more sharply drawn than in places like Cambridge, where the gulf between working-class neighborhoods and the universities was deep. O'Neill's old woman was less concerned about the war than she was with "that Harvard Square crowd." Latent divisions, wholly unrelated to the war, were opened up. The tensions which developed between students and hard-hats, parents and children were less over foreign policy than they were about domestic questions of power, class, and decorum.

Polls, with their emphatic agree/disagree focus, compounded this tragic polarization: either you are with us or you are against us. The hawk/dove model became the prevailing way in which public opinion was analyzed. Gallup reported regularly on how many people were calling themselves hawks and how many doves. Serious studies showed that Americans could not be neatly split into just two opposing camps, but the notion of a country divided down the middle had a certain dramatic appeal which put the dove/hawk kinds of questions on the front page, while more detailed analysis was buried inside.

This focus on division became a self-fulfilling prophecy. Instead of a search for common areas of agreement, the thrust of politics was divisive. "If you're not part of the solution, you're part of the problem." "America, Love it or leave it."

Gallup's question as to whether people considered themselves hawks or doves was actually a very poor test of public opinion, for it assumed that there were but two possible opinions on the war and that people held either one or the other. The final tabulations gave no indication of how many people selected one bird or the other without hesitation and how many had to be coaxed into giving a response. Nor did his figures say anything about the intensity of feelings, though an emotionally committed minority might well carry more weight than a lukewarm majority.

The dove/hawk question was not very helpful in revealing how people stood on other issues. For example, Gallup data collected in April 1968 showed that even though the various presidential candidates had markedly different views on foreign policies, their constituencies were remarkably similar in how they responded to the hawk/dove question. Forty-six percent of Wallace supporters described themselves as hawks, 33 percent as doves. For Nixon, the division was 44 percent hawks, 43 doves. Humphrey's supporters were also evenly split, 39 percent and 41 percent respectively. Eugene McCarthy, who was running on an anti-war platform, had surprisingly similar support; 37 percent of his followers described themselves as hawks and 41 percent as doves. The differences between these various percentages, with the possible exception of Wallace's, are not statistically significant. For all the attention which the hawk/dove model received, it was not very helpful in locating points of agreement and disagreement about the war.

The most important lesson which can be drawn from the polling done on Vietnam is how badly the pollsters do on issues. Their emphasis on agree/disagree questions may make for dramatic headlines, but it masks the true complexity of public opinion and, in the case of Vietnam, distorts and demeans political debate on the issue. Pollsters continue to use this format, however, because it is so much easier technically—a computer can handle favorable/unfavorable responses—and because it produces data which is journalistically provocative, even if it is not accurate.

A second lesson is our overdependence on polls. There are other indicators of public opinion. None are without flaws, and none produce the clear, crisp numbers of the pollsters, but, read together, these signs can give us a more reliable picture, albeit an impressionistic one, of the public mood. In our anxiety to understand ourselves, however, we are overly susceptible to soothsayers who can so dazzle us with technique that we lose sight of substance.

Vietnam also demonstrated how, when a battle for public opinion is being fought, the polls will be relentlessly exploited. In late 1969, Lou Harris complained in his column, "In the effort on the part of politicians to find instant approval of their positions on Vietnam, there have been some seriously exaggerated claims and equally serious under-reporting of just what current public opinion is on the war."

A recent Harris Survey had reported that 82 percent of the American people were committed to the policy of eventual withdrawal of our troops from Vietnam. At the same time, however, 59 percent of the sample qualified this commitment, supporting withdrawal only on the condition that there be self-determination for the South Vietnamese. These figures were exploited, respectively, by spokesmen for the peace movement and for the Nixon administration.

Harris was critical of both factions. "There is a tendency on the part of some excessive partisans of the peace movement to assume that when 80 percent of the public says it is 'fed up and tired of the war,' as indeed a cross-section reported in October, in turn this means that four out of five Americans want to 'bug out' of Vietnam, to use President Nixon's expression."

Harris was equally critical of the "tendency on the part of some administration spokesmen to equate opposition to the tactics of the extremists in the anti-war movement with support of the president's policies in Vietnam. 'If you don't like the Crazies, the Weathermen, and the Yippies, then it must mean that you are behind the president' is the way some of the dialogue seems to go. One needs only to recall that at a time when Lyndon Johnson was scoring a low thirty-percent confidence rating for his handling of the war, eighty-four-percent of the public also wanted to 'crack down' on anti-Vietnam protesters."

Harris was right in speaking out against this kind of propagandizing. It is only unfortunate that he and his colleagues did not repeat this criticism continually. Perhaps regular admonitions would not have kept politicians from making the claims they did, but they could have been neutralized.

More important, Gallup, Harris, and the others could have

corrected "the serious under-reporting of just what current public opinion is on the war." Much of the under-reporting was caused by their persistence in using agree/disagree questions on an issue so complex and so emotionally charged. Harris complains that he has little control over what his subscribing newspapers do with his polls. If he felt that they were being used in a misleading and irresponsible way, he should have done more than deliver an occasional sermon.

8

Who Does Speak for the Silent Majority?

George Gallup has stated, "It is my sincere belief that polls constitute the most useful instrument of democracy ever devised."

Gallup has extolled the virtues of public opinion polls throughout his long career. In a book on polls, a veritable ode, that he co-authored in 1940, he declared that they are the "pulse of democracy." Polls can tell the nation's leaders what the people really think, supposedly even better than elections can. While elections are held only once every several years, surveys can be taken around the clock so that there is a "continuous audit" of the public's mood on any issue.

"Shall the common people be free to express their basic needs and purposes," asked Gallup, "or shall they be dominated by a small ruling clique?" The pollster regarded the answer as self-evident. "In a democratic society the views of the majority must be regarded as the ultimate tribunal for social and political issues." The majority would speak, of course, through the public opinion polls.

In the first half of the 1970's we twice witnessed the dangers of government by public opinion polls. The first instance involved the deliberate attempt by some politicians to manipulate public opinion by playing on the fears and prejudices of the so-called silent majority. In the second case, Watergate, the polls helped determine the pace and course of congressional action.

The ascendancy of the "silent majority" marked the high point of the Nixon administration. Watergate, of course, signaled its fall. In both cases polls were crucial, and in both cases they were often erroneous and poorly interpreted. Rather than benefiting from the polls, as Gallup had promised, democracy suffered deeply because of them.

When President Nixon announced his plan for a gradual re-duction of American military forces in Vietnam, he claimed that his policies were supported by the "silent majority of the Ameri-can people." In many respects this constituency seemed like a reincarnation of Lyndon Johnson's old "consensus." Both were invoked to undercut the legitimacy of the "vocal minority" op-posed to United States involvement in the war.

Just as Johnson had pulled the latest surveys from his back pocket, the Nixon administration cited every available poll to prove that most Americans were hostile toward those who were demonstrating in the streets against the war. George Gallup, Jr., was recruited for a propaganda film for the United States Infor-mation Agency, in which he testified that his polls showed that the great majority of Americans stood behind Nixon's Vietnam policies.

The "silent majority" was a particularly clever concept. Con-sensus required a positive show of support; when the approval polls sunk, the consensus—and its leader—went down with them. By contrast, the leader of the silent majority does not need an affirmative vote of confidence, only the absence of overt oppo-sition. If you are not one of the 250,000 people marching in Wash-ington today, so goes the logic, then you must be supporting the president.

On the one hand, this seems to be a rather transparent device. Most Americans do not demonstrate or write letters to their congressmen—indeed, many do not even vote—but it is erro-neous to read this non-participation in the political dialogue, this silence as endorsement of the status quo. Nevertheless, the very transparency of the notion of a silent majority made it most effective politically, for it had the effect of shifting the debate away from the merits on the war issue and toward a head count of how many people were in favor of Nixon's position and how many were opposed. The substantive question—the manner of disengagement—was often lost sight of in the debate over who was on what side, which policy had the support of a majority of Americans.

The support of the silent majority had originally been invoked just to counteract criticism of administration foreign policy. In 1970 a book appeared which expanded the notion of a silent majority into the domestic arena. *The Real Majority*, by Richard Scammon and Ben Wattenberg, gave substance to Gallup's abstract ideal of government by polls and provided potent ammunition for Nixon and Agnew, as well as others, who wished to undo the progressive social reforms of the 1960's. If the Nixon administration can be credited with inventing the silent majority, it was Scammon and Wattenberg who claimed, though the magic of public opinion polls, to be able to make the silent majority speak.

The book was a success, and its impact was even greater than its general sales, for it was carefully studied by many politicians and political writers. Many readers saw it as defining imperatives which would determine the course of politics over the next decade, perhaps beyond. Because the book struck such a responsive chord, and because its thesis was fabricated almost entirely out of public opinion polls, it deserves a closer look.

The Real Majority was a survival guide for politicians, complete with strategies to follow and even speeches to give. Its language was punchy and its message direct: if liberals want to get elected, they had better scramble to the center. Scammon and Wattenberg contended that the dominant political force of our time is what they termed the Social Issue. It is not war in far-off countries which moves the voters, they said; it is not even the economy. What troubles the American voter is law and order, campus turmoil, racial unrest, and the growing permissiveness about sex and drugs. These all comprise the Social Issue.

In the authors' words, "The great majority of the voters in America are unyoung, unpoor, and unblack; they are middle-aged, middle-class, middle-minded." To survive, politicians must cater to the hopes and fears of this group. "It can be safely said that the only extreme that is attractive to the large majority of American voters is the extreme center." Scammon and Wattenberg warned that any politicians who would ignore these fundamental truths would "do so at their electoral peril."

Richard Nixon read the book three weeks before its official publication, and it reportedly convinced him that Republicans should shift their tactics for the forthcoming congressional elections. The plan had been to follow the traditional script of attacking Democrats as "big spenders," but Nixon decided that the time was right for a change. He instructed his lieutenant, Spiro Agnew, to heap blame for social disorder on the Democrats, and Agnew carried out the assignment with a vengeance.

Keeping a copy of the Scammon and Wattenberg book with him, Agnew went out on the stump for Republican candidates, railing against those who would coddle long-hairs, draft-card burners, and pornographers. He vilified the opposition as being dominated by a "little band of men guided by a policy of weakness. They plan to weaken our defenses. They vote to weaken our moral fiber. They vote to weaken the forces of law." The White House ordered *The Real Majority* in bulk. William Buckley's *National Review* warmly praised it.

But Scammon and Wattenberg were both Democrats, less interested in preaching to the converted than in getting their own party to take a turn to the right. To some degree they were successful. A few liberal newspapers and magazines dismissed the book (too lightly, as it turned out), but favorable reviews appeared in the *New Yorker* and the *Saturday Review*. The latter called it a "splendidly incisive and intelligent analysis of the American electorate."

More important, the book strongly influenced some Democratic politicians who had previously been thought of as liberal. Congressman Allard Lowenstein took to wearing an American flag pin in his lapel. Senator Adlai Stevenson of Illinois wore one too and recruited the former prosecutor of the Chicago Seven to co-chair his re-election campaign. Hubert Humphrey went him one better by getting Wattenberg himself to advise him on his 1970 campaign for the Senate. Humphrey, long the war-horse of progressive causes, whistled a different tune in 1970. Liberals had long worked for unpopular causes, he said, but now "they must show the courage to take a popular position. . . . I do not intend to let the issue of order in our society . . . be usurped by right-wingers."

Humphrey, Agnew, and all the rest were simply following the scenario which had been set forth by Scammon and Wattenberg. The authors told politicians in detail just what to say and how to say it:

> Do *not* say, "Well, I don't agree with the Students for a Democratic Society when they invade a college president's office, but I can understand their deep sense of frustration."
>
> *Do* say, "When students break laws they will be treated as lawbreakers."
>
> Do *not* say, "Crime is a complicated sociological phenomenon and we'll never be able to solve the problem until we get at the root causes of poverty and racism."
>
> *Do* say, "I am going to make our neighborhoods safe again for decent citizens of every color. I am also in favor of job training, eradication of poverty, etc., etc."

The Real Majority clearly affected the tone and character of political debate in 1970. Dozens of Democrats fell all over themselves to out-Agnew Agnew. Whether the book, and the kind of politics it spawned, actually changed the outcome of the off-elections is harder to say.

Shortly after the voting, Robert Bendiner, a member of the editorial board of the *New York Times,* wrote, "If mid-term elections may be said to prove much of anything, this one certainly proved the soundness of the Scammon-Wattenberg thesis." The election of Conservative Senator James Buckley of New York, who had run the quintessential law-and-order campaign, was commonly cited as Exhibit A in any debate about the potency of the Social Issue.

On the other hand, the Republicans, through Spiro Agnew, had exploited the Social Issue to the hilt, yet they failed to increase their numbers in the Senate, even though it had been predicted that they had a real chance to get control for the first time in almost two decades. Scammon and Wattenberg counter with the argument that the Republicans were thwarted only because enough Democrats heeded their book and fought fire with fire.

The Real Majority could not have been published at a more propitious time. We were still mired in a purposeless and debilitating war abroad, and at home demonstrations were becoming ever more violent. Crime was on the rise, and youth had apparently embraced an alien culture. Scammon and Wattenberg appeared in the midst of this scene and confidently offered a cure, even before we were certain just what it was that ailed us. Their tone was direct and their answers authoritative. Many politicians found the combination irresistible.

Few looked below the surface to see that *The Real Majority* was a book flawed in its assumptions and sinister in its implications. The book in fact cynically exploited the polls to clothe the authors' personal political dogma in the sacred robes of public opinion.

Although the authors righteously disclaimed the connection, there is in the end little distinction between saying that politicians must recognize that the country is "unyoung, unpoor, and unblack" and advocating that politicians be anti-young, anti-poor, and anti-black.

Only a few commentators saw that behind the authors' glib list of political dos and don'ts was the idea that we should turn our backs on certain fundamental values. Michael Janeway of the *Atlantic* observed: "Something gets a little lost in the shuffling around of these straw men. What was it, freedom of speech, or of the press; or the right of the people peaceably to assemble, and to petition the government for a redress of grievances? . . . The authors don't say that liberals should run out on the Bill of Rights; just stop sounding like . . . Bill-of-Rights liberals."

The shallowness of Scammon and Wattenberg's analysis and the dangerous implications of their thesis would have been more readily apparent had not the authors been astute enough to dress up their book with opinion polls. It was a most effective gambit. The relatively few people, like Janeway, who took the book on, came off as being anti-democratic. It was not Scammon and Wattenberg speaking, it was the people, expressing themselves through the polls.

For the proposition that the country is "unyoung," for example, the authors cited a Gallup Poll which reported that Americans opposed legalization of the use of marijuana by 84 to 4 percent. As proof that America is "unpoor" and "unblack," they pointed to another Gallup survey showing 90 percent of the population opposed to reparations for past racial injustices. The authors bombarded many of their readers into agreement with page after page of such polls.

Public opinion polls, said the authors, are of "immense value," not acknowledging either the technical pitfalls of polling or the subjective nature of any analysis. When the book came out some critics noted this over-reliance on polls, but no one took on Scammon and Wattenberg on their own terms. No one spoke up to suggest that perhaps the polls do not really say what Scammon and Wattenberg said they did.

In a later book, Wattenberg curiously let slip the remark, "one can defend almost any position on almost any subject by referring to public opinion polls." That off-hand admission is the key to Scammon and Wattenberg's work. The authors ransacked the vast bank of polling data to find results compatible with their own point of view. In this cynical and self-serving process, they simply discarded all the evidence which cut the other way.

Writing several years later, Louis Harris forcefully illustrated with his own survey results just how one-sided the Scammon and Wattenberg presentation was. His polls indeed did show a great majority against "legalizing the sale and use of marijuana," but can America really be called "unyoung," asked Harris, when, by a margin of 72 percent to 20, young people are praised for "being more honest and open about sex."

True, 88 percent of the people Harris interviewed wanted to "make people on welfare go to work," but that does not mean America is "unpoor," for in response to another question, 87 percent said they favored "a federal program to give productive jobs to the unemployed."

Looked at in isolation, a Harris Survey which showed 81 percent of the population opposed to compulsory "school busing to achieve racial balance," makes America seem "unblack," but such a conclusion ignores still another Harris Survey which showed a substantial majority still supporting "desegregation of the public schools."

This kind of selective use of polls was the hallmark of Scammon and Wattenberg's book. The authors, who stand in the right wing of the Democratic Party, tried to use the polls to prove that they were smack in the middle of the real majority. Abbie Hoffman, with a little imagination and certainly more humor, could have used the polls which Scammon and Wattenberg threw away to prove that he too is a middle-American at heart.

If the Social Issue was a fabrication and the silent majority a self-serving cliché, what happened to the politicians who followed Scammon and Wattenberg's advice? A clue to the answer can be found in a series of Yankelovich surveys taken in New York State during the 1970 campaign. Yankelovich reported that concern about "law and order" did indeed grow throughout the campaign. But was this really a victory for the politicians who were playing on this concern and, in fact, were probably responsible for its growth? The Yankelovich study showed clearly that it was not. Spiro Agnew had become the outspoken champion of law and order. In September, when he started out on the stump, people had said by a 39 to 30 percent margin that Agnew was "helping the United States," but by the end of the campaign a significant switch had taken place. By a 46 to 34 percent margin people now said he was "hurting the United States."

At the time, there was little notice given Yankelovich's results, but the message was clear: rather than providing the key to electoral salvation, a politician who recited the Scammon and Wattenberg catechism risked damnation by the voters. Its shallowness, its emphasis on perception over substance, spelled its own undoing.

That this had to be true becomes clear when one examines the instructions which Scammon and Wattenberg gave the Democrats for dealing with the Social Issue in 1972: Do not offer new policies or revitalized values, simply mouth new words. "To begin, it should be clear to any practicing political type that the keynote address at the 1972 Democratic Convention will include a passage to this effect: 'Fellow Americans, in the past four years crime in America has gone up by ____, while the population has increased by only ____%.' Of course, the same paragraph was recited several hundred times by Candidate Nixon in 1968, who pointed the accusing finger at eight years of Democratic rule. In

1972, however, the accusing finger will be pointed toward four years of *Republican* rule."

Exploit the fear. Blame the other guy. If the public is distressed about the unraveling of the social fabric, then cater to that concern even if by doing so you make it unravel further. If Gallup says that people are worried about crime, speak not of the rights of the accused or police brutality. Just promise a crackdown on the criminal element.

For all the respect for the common people that those who practiced this kind of politics claimed to have, they gave them precious little credit for being able to see through this sort of pandering. People were concerned, but they wanted answers, not empty words. Americans were worried about a society which was coming unglued and for that very reason ultimately rejected politicians who played on division.

Scammon and Wattenberg's list of political dos and don'ts could at best buy a politician a little time. Just as we came to pay for the unfulfilled promises of social equality in the 1960's, so the law-and-order line set up assurances which, if not kept, were bound to be thrown back in the faces of the politicians who uttered them. Those who vow to put an end to disorder in society will get called on their words if there is no change. Change cannot come about, however, when politics is just a process of manipulating images and slogans.

When Richard Ottinger announced his 1972 bid for a congressional seat from New York, he sternly promised, "You're going to see a law-and-order Ottinger." Bowing and scraping to the Social Issue, he criticized a "society that has become too soft on crime, agonizing more over the fate of criminals than over the fate of victims and potential victims." Voters who listened to Ottinger apparently thought his words were hollow. He had, after all, been in Congress for three terms. What had he done about crime then? Ottinger lost the election.

The fundamental weakness of the Scammon-Wattenberg thesis, the reason it carried the seeds of its own destruction, was that even though the authors extolled the importance of public opinion, they really underestimated the intelligence and sensitivity of the American voter. They overstated people's susceptibility to emotional appeals and discounted their ability to spot a phony.

As it happened, there was another reason why the Silent Majority, the Social Issue, and the Real Majority all went down the drain so quickly. In the end, those who most vigorously exploited this sort of politics, namely Nixon and Agnew, were themselves consumed by it. When they fell, their politics toppled with them.

Writing in 1971, Scammon and Wattenberg concluded that the Social Issue was then essentially a pro-Republican issue. "Neither Muskie, Humphrey, Kennedy, McGovern, Harold Hughes or Ramsey Clark will get very far calling Nixon, Agnew and Mitchell soft and permissive." Other commentators might then have made the same observation, but unanticipated events in the next several years showed how wrong that conclusion was.

While denouncing "crime in the streets," Spiro Agnew was shaking down construction firms. John Mitchell corralled thousands of anti-war demonstrators into a football stadium used as a makeshift jail. Three years later he himself faced prison. At the top of course was Richard Nixon, who orchestrated what truly can be called the political crime of the century. The downfall of Nixon, Mitchell, and Agnew was also the undoing of exploitation of the Social Issue, for they, who had been the loudest in speaking out against crime in the streets, in the end were seen to have been the biggest criminals. Exploiting fears and practicing the politics of expediency, they lacked an ideology to set them straight.

As late as 1974 Wattenberg refused to acknowledge that Watergate could leave a lasting mark on American attitudes. To do so would be to admit that our attention had been distracted from the Social Issue. "Watergate is essentially ephemeral. It is perceived to involve a series of ugly events, not an ugly ongoing condition."

To prove that the issue really was transitory, Wattenberg again turned to public opinion polls, this time citing a Gallup survey which indicated that of the ten most admired men in America, eight were in political life. This is a rather slender reed on which to hang the claim that Americans still revere politicians. When Gallup tried to make his annual most-admired man poll in 1974, the people who were interviewed were unable to come up with anybody they admired! Gallup ultimately had to alter the question by presenting a list of possible candidates for the title.

Polls can, in fact, show us how deeply Americans did feel the ugliness of Watergate. They also show how the "silent majority," if indeed there ever was one, eventually turned against the Nixon administration. Unfortunately, many of the polls were poorly drafted, and even those which were better constructed were often misinterpreted. As a consequence, the true state of public opinion on Watergate was obscured by the polls.

Politicians and commentators who followed the published surveys on Watergate read them as showing that even though most Americans believed almost from the outset that Nixon and others in the White House were implicated in the Watergate bugging and break-in, either in the planning or in the cover-up, only a minority supported impeachment. Peter Rodino, chairman of the House Judiciary Committee, thinks that it was congressional leadership which brought public opinion around. "Very early on the polls showed people were reluctant to support impeachment. Only after the Judiciary Committee and the courts began to act did the polls show people supporting impeachment; even then, they did not support actual removal of the president."

The published polls seemed to conform to this pattern. In December 1973, for example, Gallup reported that 76 percent believed that Nixon was involved in the Watergate scandal in some capacity, yet only 35 percent favored his removal from office. The conventional explanation for this apparent dichotomy in attitudes is that most Americans regarded Watergate as "politics as usual." Indeed, for a long time this was the principal defense of the Nixon administration: they had done nothing worse than the Democrats had done in other elections.

Politicians who are nervous about public opinion read polls like Gallup's as warnings to go slowly on the issue of impeachment, lest they risk holding an unpopular position. Even those who saw that much was at stake felt that they had the problem of educating Americans about the seriousness of the offenses before the Congress could initiate impeachment proceedings. Yet this was a dangerous misreading of the public mind, a misreading which probably kept Nixon in office longer than necessary and might have enabled him to complete his term.

People were reluctant to answer yes when pollsters asked whether the president should be impeached less because of cynicism or a feeling that the issues were not important than because most people did not understand what impeachment really is.

In late 1973 Burns Roper took a poll which tried to probe attitudes more deeply than Gallup and Harris had done in their published surveys. Roper asked those who were opposed to impeachment, "Is that *more* because you don't believe the charges against President Nixon are true, or *more* because you don't believe the charges are serious enough, or *more* because you think it would be too destructive to the country?"

The question was not worded well, as it presumed that there were only three reasons why a person might oppose impeachment and that he would have to favor one over the other two, but at least it provided a bit more information. Fifty-three percent answered this forced-choice question by saying that they were principally worried about the effect on the country.

Much more important, however, was Roper's concurrent discovery that apparently fewer than 50 percent of the people he polled knew what impeachment was! A majority thought that impeachment meant forced removal, not simply investigation and indictment by the House of Representatives. When this majority was asked whether they favored impeachment, they believed that they were being asked to pass final judgment before there had been a full hearing. Little wonder, then, that people suspected the worst, but would not pronounce sentence. In saying that Nixon should not be impeached, many really meant only that he should not be denied a fair trial.

What is astounding is that in late 1973 so many people could still have been ignorant about impeachment. That summer John Dean had told all to Sam Ervin's Watergate committee. The existence of the tapes had been discovered, then the suspicious gaps had been revealed. Elliot Richardson and William Ruckleshaus had been forced to resign, and Archibald Cox had been fired. Peter Rodino was laying the groundwork for the House Judiciary Committee investigation. In spite of the fact that Watergate had dominated the news for a year, most Americans still could not properly define impeachment.

What is even more astounding is that this did not stop Gallup and Harris from asking everyone they interviewed whether or not they favored impeachment. The responses of those who understood the process were lumped together with the answers of the large number of people who did not.

Indeed, there is reason to believe that some of the pollsters themselves did not understand the impeachment process, and that their misunderstanding was reflected in their questions. Throughout 1973, for instance, Gallup's basic question was, "Do you think President Nixon should be impeached and compelled to leave the presidency, or not?" Well into 1974 Harris continued to ask people whether they wanted Nixon "impeached and removed from office." The pollsters seemed to regard impeachment and removal as one and the same thing. It was as if they had asked, "Do you think Nixon should be tried and hanged?"

In contrast with Gallup, Harris is known as one of the most sophisticated draftsmen of questions in the polling business. It is hard, therefore, to believe that his formulation of the impeachment question was just an oversight on his part. Indeed, Harris' clandestine association with Colson and others in the White House strengthens the suspicion that the wording was carefully chosen so as to dampen anti-Nixon responses.

Pollsters, particularly those who, like Gallup, believe that a poll is a referendum in which people can cast a vote for or against a particular proposition, all too frequently fail to look beyond what people say to see what they really mean. Gallup is a direct man, and his style permeates the work of his organization: if you want to know how people feel about impeachment, you simply ask whether they are in favor of it or are opposed. Gallup should have tested public knowledge of the process. The big story really was that people still did not understand that impeachment is only a prelude to a full trial by the Senate. Responses based on such a fundamental misconception really are meaningless.

The few pollsters who had the imagination to look deeper found some important results. Pat Caddell, who had conducted a number of private surveys which had touched on the issue, discovered just how critical the wording was. "The problem really was semantic. The way that Gallup posed the question, it seemed that only 30 percent said Nixon should be impeached. But if you asked, 'Do you think the president should be tried and removed from office if found guilty', then 57 percent said he should." At an early stage, then, Americans actually favored the principle of impeachment, even if they did not understand the meaning of the word itself.

As a consequence, the Gallup and Harris polls had made the

American people seem much more hesitant about having the full Watergate issue aired and resolved than was really the case. It was only a small difference in phrasing, but had the best-known pollsters worded the questions differently, it would have been clear much sooner that Americans did not want to sweep Watergate under the rug.

Pat Caddell, who in 1973 and 1974 was doing a lot of polling work for Democratic senators and congressmen, is convinced that the nationally published polls had a stagnating effect on the impeachment process. As long as the polls seemed to show a majority of Americans opposed to impeachment, most congressmen were not going to push it; and as long as politicians did not speak out, their constituents continued to tell Gallup and Harris that they were against "impeachment and removal." Caddell concludes, "It was a very dangerous situation." He believes that events like the firing of Cox and the apparent tampering with the tapes were so shocking that people ultimately answered yes to Gallup's and Harris' questions in spite of the way they were written. Only then, when public opinion was far ahead, did the politicians dare to take a stand.

Caddell is probably right. Because of their poor wording, the national polls on impeachment lagged well behind actual public opinion, and as a result politicians were overly cautious in speaking out. At the same time, however, there was some behind-the-scenes politicking involving polls.

In spite of his White House connections, Lou Harris actually worked both sides of the street. In the spring of 1973 Congressman John Rhodes summoned him to meet with the Republican leadership to discuss the implications of Watergate. At that early date, Harris warned the Republicans that their party could be badly hurt if it fought to keep Nixon in office. Yet at the same time as he was giving this private warning, his published polls were indicating that a majority opposed impeachment. Those who did not speak with Harris firsthand quite reasonably believed that his polls meant that impeachment should not be pushed.

The Democrats were talking to pollsters too. House Majority Leader Tip O'Neill cornered his colleagues, one by one, to show them a poll taken by Bill Hamilton. In essence, the survey was an attempt to test whether it would make a difference in the 1974 congressional races if a candidate were considered pro-Nixon or anti-Nixon. According to O'Neill, the results indicated that, whether they were registered Democrats or Republicans, far fewer people were inclined to vote straight party lines. The poll showed that there was a great deal of softness among Republican voters, so that even Republican congressmen who had held their seats for many terms might be turned out of office if a Democratic challenger made the election a referendum on Nixon.

O'Neill thinks the poll helped him get some Democrats who had been sitting still to back impeachment efforts, and, just as important, it neutralized some potential Republican opposition. In the end, the off-year election may have borne out the Hamilton poll. Not only were there substantial Democratic gains, but some of those Republicans on the House Judiciary Committee who until the end had been most vocal in their defense of Nixon made the ultimate political sacrifice the following November.

The White House, in turn, relied heavily on polls in its fight to hold on to power. During 1972, the campaign year, the administration had the benefit of the information spewed out by the million-dollar polling operation run by Robert Teeter for the Committee to Re-Elect the President. Thereafter, Colson was in steady contact with Albert Sindlinger who could report the daily ups and downs of the president's stock. Colson also continued to monitor the published polls, and confer with the men who took them, in order to fashion a counteroffensive on the public relations front.

Much of the White House's strategy in 1973 seemed to be an attempt to ride out the issue. "One year of Watergate is enough," said the president, hoping to tap the nerves of those people who told pollsters they thought the press was being unfair. Indeed, in June 1973 Gallup had reported that 44 percent believed there was too much Watergate news. The White House shut its eyes, however, to a poll a month later by Harris which reported that, 51 percent to 37, Americans wanted Richard Nixon to appear before Senator Ervin's committee.

In an administration that was so conscious of the polls, it is hard to imagine that any major decision would be made without considering its impact on public opinion. So long as Gallup and

Harris showed a majority against impeachment, Congress might not act. Buying a little time might cause the Watergate scandal to run out of steam. But how was the White House to buy that time?

In late October 1973, Nixon pressed for the firing of Special Prosecutor Archibald Cox. Attorney General Elliot Richardson had made it clear long in advance that if Cox went he would go too. In forcing the issue, Nixon made a dreadful miscalculation about the public response to his action. Up until then it had been almost an iron-clad rule that whenever a president did anything decisive, he would rally public support at least for a time. That was true in Vietnam for both Johnson and Nixon; whether it was an increase in the bombing or a halt, it would produce a spurt in the public's approval of the president. It was also true for Kennedy and Eisenhower, even when they blundered badly.

This pattern, which has been so clear over the past twenty years, was known by the pollsters who were close to Nixon, and Nixon knew it himself. In his book *Six Crises,* he wrote, "People invariably rally around a president in a period of international crisis." Did he think that would apply just as well in a domestic crisis, when a president purged people from the government? Did Nixon believe that the Saturday night massacre could be his Bay of Pigs, his U-2? Surely his action would spark criticism from the liberal establishment, but they had jumped all over Ike and JFK. The people had rallied around the president before; they would do it again! Americans do not like to see their presidents humiliated, even when they are wrong.

If this was Nixon's reasoning (and it is hard to believe it could have been otherwise), he must have been shocked at the massive reaction against his firing of Cox. The events took place on Saturday night, when the thoughts of most people were far from politics, yet the outpouring of telegrams, phone calls, and letters to Congress and the White House itself was unprecedented in our political history.

Even those who were anti-administration were stunned at both the number of messages—literally millions—which were sent, and the intensity of the feelings they expressed. Gallup reported that Nixon's rating had fallen below 30 percent for the first time. Newspapers which were once friendly to Nixon now called for his resignation. Columnist William Buckley said he expected that Nixon would resign at the urging of conservative Republican senators like Barry Goldwater.

Nixon was in the midst of a crisis which could have ended his term then and there. Even those who were still not prepared to make a judgment on the Watergate question wondered out loud how Nixon could continue to govern with so little support. Instead of buying more time, Nixon seemed to have spent it all. How could he recoup?

Five days after the Cox firing, the attendant resignations, and the resulting public outcry, Secretary of State Kissinger announced that the United States military forces had been placed on high alert around the world. Tensions were rising in the Middle East at the time, but no reason was given for this action, even though such an alert had not been ordered since the Cuban missile crisis.

Commentators openly speculated that the military action was a diversionary tactic meant to draw the public's attention away from Watergate, and perhaps to underscore the need for a strong president in perilous times. Kissinger flatly denied these speculations. "It is a symptom of what is happening to our country that it could even be suggested the United States would alert its armed forces for domestic reasons." He gravely vowed that when the reasons for the alert were made public, which he promised would be in several weeks, everyone would see that the action had been necessary. Many people remained skeptical, but the fact that United States planes with armed nuclear devices were circling in the skies did push much Watergate news off the front page.

Perhaps the global alert was wholly unrelated to Watergate. Perhaps there were serious threats to our security which warranted an overt show of our preparedness. But if that is the case, why did Kissinger fail to keep his promise to justify the incident? Whatever its genesis, it did have the effect of staying Nixon's execution, if only temporarily.

The polls apparently convinced the Nixon people that Watergate was an image problem, a problem which Americans would tire of if it dragged on long enough. They hoped, futilely, for a backlash against the press. They tried to argue that nothing worse than politics as usual had happened. But as the evidence piled up, Nixon was increasingly backed into a corner from which he could not escape. Had he recognized the substance of the issues which were at stake, and had he been more forthright in meeting them, it is likely that he would have served out his term, bruised, no doubt, but still president.

In his book *Breach of Faith*, Theodore White quotes former Secretary of Health, Education, and Welfare Caspar Weinberger on how the White House misperceived the public concern about the Watergate scandal. Weinberger said: "I don't think Nixon began to take the matter seriously until the polls went down; then it became more than a PR matter. But you see, they had no philosophical rudder to their administration to take them from there. They couldn't see the matter morally."

Instead, the Nixon administration generalized from the Scammon and Wattenberg thesis and dealt with perceptions, not realities and values. Operation Candor was an effort to appear honest, not to be honest. Had Gallup and Harris asked the right questions, it would have been apparent earlier that public relations was not enough.

The nationally published polls helped prolong Nixon's tenure, but in the end they helped to terminate it. By mid-1974 a majority of Americans were able to say, "Yes, the president should be impeached." If Congress could not move quickly enough to pass judgment on Nixon, then the people would. Just as the polls in 1973 had been read as a vote of confidence in him, so they were seen in 1974 as proof that he had to go. It was not just the liberal establishment speaking; according to Gallup's figures, in every part of the country those who wanted Nixon out of office outnumbered those who wanted him to stay.

Louis Harris says: "There is no doubt in my mind that our polls on impeachment of Richard Nixon had a profound effect. I know they did." Harris thinks that congressmen who talked privately with him and other pollsters were swayed by what they heard. He also thinks that the published polls had an enormous impact. "I know that the *Chicago Tribune* put our final pre-resignation poll in a six-inch banner headline. The fact that 63 percent of the people in our poll favored impeachment had a profound effect on this decision, because the *Chicago Tribune* putting it on the front page is a lot different from the *New York Times.*"

There was concern among some pollsters about whether they should be polling while Congress was considering the impeachment of a president. Columnist David Broder, writing in April 1974, said that "some of the best-known of the pollsters are asking for advice on their role during the period when judgment is being passed on the president." According to Broder, these pollsters were aware that they would be under heavy pressure from their client newspapers to report every facet of the impeachment process. Just as they did during the Senate Watergate hearings under Chairman Sam Ervin, the polls could report on which witnesses were believed and which were not, and how the public was responding to the performance of each of the congressmen.

Most worrisome to Broder and the pollsters with whom he conferred was the possibility that the polls could become a national referendum by which the guilt or innocence of Richard Nixon would be determined. "What worries them—as it ought to—is whether it is proper to unleash this wealth of information on the men and women who must ultimately decide the president's fate or whether this kind of reporting may reduce the impeachment and trial of Mr. Nixon to the level of mob rule symbolized by the Roman populace signaling the fate of a gladiator by pointing thumbs up or thumbs down."

If, however, awareness of their vast and dangerous power gave the pollsters some second thoughts, it did not in the end stop them from polling. Perhaps the fact that both Gallup and Harris only scratched the surface in their polls may reflect some personal reservations on their part about what they were doing. But the resolution of their doubts via publication of incomplete and misleading information was certainly worse than not publishing at all or going ahead with intensive studies.

Nixon's resignation kept us from seeing what role public opinion in general and the polls in particular would have had in the full impeachment process, but a number of congressmen gave serious thought to the weight they would give to their constituents' views.

Peter Rodino, chairman of the House Judiciary Committee, which ultimately voted to recommend adoption of articles of impeachment, says that his committee gave the polls relatively little attention. "Those of us who had to make the decision had an awesome task. The country was in a crisis. While some may have viewed the polls to get a reading, in the end what everyone did was to base their judgment on the facts and the best interest of the country." Rodino says that his fellow members of the committee shared a feeling that the only other impeachment in our history, that of Andrew Johnson, had been thoroughly partisan and irresponsible, and they did not want their actions remembered the same way.

Charles Wiggins, one of the most influential Republicans on the committee, says that he shared Broder's concern that the polls could contribute to mob rule, but he regarded it as a problem without a clear solution. "I personally tried to insulate myself from all extraneous forces, including my constituent mail on the subject. I didn't read any of that from the first of the year on. I couldn't protect myself from the polls in the *Washington Post* and the *Los Angeles Times*, but I tried to discipline myself not to be influenced in any way by the popularity of President Nixon."

Wiggins essentially sequestered himself from public opinion during the Judiciary Committee hearings, but had the matter gone to the full House he would then have regarded it as one factor that must enter into the impeachment decision. Wiggins regarded impeachment as a two-stage proceeding. In the first stage the House Judiciary Committee considers the applicable facts and law to determine whether, in Wiggin's words, "the president is a proper candidate for impeachment." Wiggins and others on the committee believed this was a uniquely legal proposition, a careful weighing of evidence and law. Wiggins says, "I think I would have felt unclean if I had listened to my constituency for the resolution of legal issues."

Had the articles been considered by the House and Senate, however, Wiggins believed that public opinion could have been properly heeded. "When the issue reaches the House and Senate, political considerations, with a capital *P*, are proper. It's appropriate for both the House and the Senate to decide whether it is in the national interest that a given president *should* be impeached, as distinguished from whether he *can* be." Thus, in Wiggins' view, a president might indeed be guilty of "high crimes and misdemeanors" but still not be expelled from the White House if public opinion was such that it was in the national interest for him to remain. Such might be the case when a president has committed a criminal offense wholly unrelated to his conduct in office.

It is thus possible that public opinion polls could have had an ever larger role in the impeachment of Richard Nixon. Congressmen who felt compelled to ignore them while serving on the House Judiciary Committee would have consulted them if the question had gone before the full House.

Perhaps we should devise a system in which the people can express their will in such situations, but to allow the pollsters to speak for the people would be a disaster. The best-known pollsters bungled the impeachment question in 1973 and 1974. We can only hope that, should the situation arise again, it will not be left to a Gallup or a Harris to determine whether a president stands or falls.

9

Political
Poll-Vaulting

When former Attorney General Ramsey Clark ran for the Senate from New York in 1974, he broke a cardinal rule of modern campaigning. John Martilla, a professional political consultant, had urged Clark to undertake an intensive poll, at a cost of sixteen thousand dollars, to "establish a benchmark" against which later progress could be measured. Clark rejected the idea as costly and useless; during the campaign a half-dozen other firms solicited his polling business, and he turned them all down. He ran without taking any polls, professional or otherwise. He also lost.

Starting in 1972, Ohio Governor John Gilligan hired Washington, D.C., pollster Peter Hart to do a series of studies in anticipation of his 1974 campaign for re-election. Gilligan, who had frequently been mentioned as a future contender for national office, invested seventy thousand dollars for five different polls over two years. Six months before the 1974 election, however, he stopped polling, apparently confident that he was a sure winner. Like Ramsey Clark, Gilligan also was beaten.

The experience of Clark and Gilligan was not lost on other politicians. If you want to be elected today, you start polling early and you keep polling up until the last minute. Having a prominent pollster is a mark of political status that establishes a candidate's legitimacy. Early in the 1976 presidential campaign, several thinly financed Democrats were claiming that they had hired pollsters when that really was not the case. Pollster Peter Hart says: "I think it's really a defensive thing. They're not really doing the polling, and that's a source of embarrassment. If the press ever calls them on it, they just say that their polling operation is secret."

Though political polling was done sporadically in earlier elections, it did not really begin to flourish until the last twenty years. In that brief period it has become an integral part of the campaign process, touching every phase of electioneering.

Before a prospective candidate can get the political and financial support necessary to launch a campaign, he must have a poll which shows that he has a respectable chance of winning. Favorable polls are essential for political fund-raising, and, as California Democrat Jesse Unruh once put it, "money is the mother's milk of politics." As early as the nineteen-thirties Will Rogers remarked, "Politics has got so expensive that it takes lots of money even to get beat with," and that is all the more true today. A candidate who can flash a good poll will get contributions, even grudging ones, since people do not want to alienate those who are going to be in power. A bad poll, however, can discourage a candidate's best friends. Who, after all, wants to throw good money away?

George McGovern is not sure that the polls published during the 1972 presidential campaign had any direct effect on the voters, but he is positive that they had an impact on important people who had bankrolled the Democratic Party in the past. "You see it with certain people who are looking for favors from the government. I know some people personally who had no regard for Nixon and they didn't prefer his ideology, but they gave money and signed ads. I know them well enough to know why they did it. The polls said that Nixon was going to be a big winner, and they didn't want to get frozen out."

Most pollsters concede that success or failure in early polls, often taken a year or more before an election, can determine whether a candidacy lives or dies. These early polls have little to do with a candidate's true vote-pulling power, yet political financiers give them great weight. California poll-taker Mervin Field says: "There's a lot of stupidity among those who make major contributions to candidates. You wonder how they were able to make their money in the first place." The early polls are meaningless in their own right, but because the money-men respond to them, they can easily become self-fulfilling prophecies.

Political surveys are also a very effective means of collecting another kind of electoral currency, endorsements. Long before the 1972 presidential campaign began in earnest, Senator Edmund Muskie was able to line up support from prominent Democrats in Congress and state houses throughout the country, largely on the strength of polls which showed him to be both the preferred candidate in his party and the strongest challenger to Richard Nixon. Muskie's campaign ran out of gas, however, and many of the party regulars who had agreed to run on slates pledged to him found themselves locked out of the convention.

Similarly, in 1970 former Supreme Court Justice Arthur Goldberg was able to capture the Democratic nomination for the governorship of New York by flashing an Oliver Quayle poll which showed that he could beat the incumbent, Nelson Rockefeller, by twenty-five points. Quayle, who was well respected in Democratic circles, reported that Goldberg was the only candidate who could beat Rockefeller, and that he would be a "tremendous asset" to the rest of the ticket. Party leaders were reluctant to throw their support to a newcomer, but Quayle's report was so enthusiastic that they felt they had little choice.

Even with the backing of the Democratic machine, however, Goldberg proved to be an abysmally inept campaigner. He just barely beat Howard Samuels in the primary and went on to get trounced by Rockefeller in the final election.

Muskie and Goldberg essentially practiced a political ploy which had been perfected ten years earlier by John Kennedy. Kennedy, of course, was able to carry it much further. The Kennedy people were experts at using public opinion polls to prod, or club, delegates into jumping on the bandwagon before it pulled out. In his account of the Kennedy years, former presidential aide Theodore Sorensen wrote, "Pains were taken to make certain that all publicly conducted polls, by Gallup and others, were properly taken into account by those we hoped to sway."

The process, however, was not nearly as delicate as Sorensen's language suggests. There was a strongly felt ethic among the Kennedy people that "you're either with us or against us." Democrats who were less than enthusiastic about Kennedy were shown polls which in essence said that even if they did not come aboard Kennedy would win, and they would be on the outside looking in.

The principal task of Louis Harris, Kennedy's personal poll-ster, was to churn out polls which would keep the bandwagon rolling. Sorensen describes one aspect of Harris' work. "Equally important, however, were the results of privately financed and conducted polls which were primarily for the senator's informa-tion though given to friendly politicians and columnists."

The late Elmo Roper, who apparently never missed a chance to criticize Harris, his former employee, was more explicit about just how these private polls were used. He bitterly attacked those "so-called public opinion researchers," meaning Harris, who al-low their polls to be exploited "rather openly for propaganda purposes." Roper recounted an incident which shows just how willing Harris was to have polls used this way. "In one case polls were even leaked to an opponent showing he was seven points ahead. When he called me up long distance to ask me why I thought he would be shown a poll that Mr. Kennedy had paid for and that Mr. Kennedy had not yet seen, I said, 'Well, has it occurred to you that they would like to see you let up on your campaigning a bit?"

Hubert Humphrey, the victim of much of this pollsmanship, today recalls Kennedy's and Harris' work with more envy than resentment. "Kennedy used the polls with a master's touch. Lou Harris was doing a lot of it for him. The polls were always being planted in the newspaper columns, Scotty Reston's and every-body else's. The information was made available; here it is, here's how our man is doing. When you're running high like that, and hard, you get the reputation as a front-runner."

A poll which finds its way into James Reston's column acquires an almost irrefutable legitimacy, especially among those who make contributions or give endorsements. It also colors the way in which other reporters cover the campaign. The political fall-out can be tremendous.

Humphrey had to contend with this in the spring of 1960; Nixon had to face it in the fall. Nixon's account of the 1960 campaign in *Six Crises* is written in a carefully modulated tone; nevertheless, it is clear that Nixon felt bitter about the way Harris' polls were used against him. He specifically describes the "blitz" of the last few weeks, when Harris was telling people in the press that his polls showed a Kennedy margin of four or five million votes. That estimate contradicted Nixon's own surveys, as well as some of the published polls, but many commentators believed it. Nixon feels that the Harris leaks had a demoralizing effect on his own campaign and caused some political writers to conclude that Kennedy had the election locked up.

It is one thing to leak favorable polls; it is another to have them believed. Harris and Kennedy were very successful at both.

In many cases political pollsters do much more than just generate numbers for the propaganda machine. All campaigns involve the management of scarce resources—money, workers, and the candidate's time—and most of the important decisions involve how these resources can best be used.

Pat Caddell's surveys for George McGovern in 1972 were used to determine how to get the most mileage out of every campaign appearance and each advertising dollar. In June, for example, McGovern had to decide whether to campaign in California right up to the day of that state's primary or leave early to appear at a governors' conference in Houston, where he hoped to make peace with Democratic Party regulars. He ultimately made the trip, but only after Caddell's studies showed he could leave California without risking losing the primary.

Political polls are used to discover who the voters are and how they feel, so the candidate can plan his strategy accordingly. Campaigning hard where the voters are already strongly favorable is as unproductive as working those areas where people are decidedly hostile. Some pollsters are better than others at providing this kind of guidance. In 1960, for example, the Kennedy team grew increasingly disenchanted with Lou Harris' advice on this score. In several instances it was felt that Harris had induced Kennedy to campaign in areas where he had nothing to gain at the cost of ignoring other places where the election was in doubt.

Other politicians, however, have been delighted with the strategic advice their pollsters have given them. Those in charge of the re-election campaign of Senator Warren Magnuson of Washington were impressed by a Peter Hart survey which indicated that Magnuson drew significantly more support from voters who had never seen him personally than from those who had. The senator had a reputation as a dedicated and effective public official, but in person he tended to come off as dull and bumbling.

The managers felt that their candidate had to make some personal appearances or risk the charge that he was running an absentee campaign. So that his personal exposure would not be too costly, they finally decided to have Magnuson stump primarily in areas that he had lost heavily in the past! The strategy worked; Magnuson was easily re-elected, and those involved with the campaign gave great credit to Hart's polls.

Television has largely supplanted street-corner rallies as the best means of reaching voters, and here too the influence of political pollsters is felt. House Majority Leader Tip O'Neill recalls a Massachusetts campaign that he managed in 1962 in which polls guided critical campaign decisions about the use of television. O'Neill was running the campaign of Endicott Peabody, who was challenging Republican Governor John Volpe. Surveys taken by Joe Napolitan a day or two before the election gave Volpe a slim one- or two-point lead.

O'Neill analyzed the results from different areas of the state and saw that certain Catholic neighborhoods which had traditionally gone heavily Democratic were giving Volpe substantial support. In O'Neill's words: "Our polls showed that a lot of people in the Irish Catholic areas of Boston would rather vote for an Italian Knight of Malta than a son of an Episcopal bishop, even though he was a Democrat. As a consequence, we put together a television program for Monday night where we had Robert Kennedy, John McCormack, all kinds of Irish Catholics in public life, speaking up for Peabody."

As the returns came in, it became clear that this last-minute appeal had worked. Peabody carried the key areas strongly, and that gave him enough of an edge to win the election overall by several thousand votes. O'Neill says, "There is no doubt in my mind that Joe Napolitan's polls really made the difference."

Polls are used to determine not only where television appeals should be pitched, but also what kind of image the candidate should project through his advertising. Surveys conducted for John Lindsay in 1969 when he was running for re-election as mayor of New York indicated that voters would have been willing to forgive him for his mistakes if he had been frank about acknowledging them. Instead of trying to build a campaign based on a four-year record of unblemished achievement, the Lindsay staff pushed the notion that the man had done as well as could be expected. The tone of the advertising was epitomized by the Lindsay slogan, "It's the second toughest job in America." Lindsay was re-elected, beating Mario Procaccino by a slim four points. Richard Aurelio, Lindsay's campaign manager, said after the victory, "We couldn't have won without the polls."

Though political pollsters freely admit their influence in strategy decisions and image-making, few openly concede that their surveys are also used to mold their candidates' views. Privately, however, some take credit not just for how their clients appear but for what they say. During the fall of 1974, for example, Pat Caddell was polling for Michael Dukakis, who was running for the Massachusetts governorship. Caddell bragged to some people that his polls had convinced Dukakis to backtrack from his previously strong stand in favor of court-ordered busing.

In short, pollsters have invaded every aspect of the election process, from beginning to end. As West Coast pollster Don Muchmore has said, the pollster "has become at least an equal partner with the campaign manager and, in some instances, even a 'super campaign manager' who determines the direction of a campaign." According to Muchmore, "Today's pollster is able to alter the outcome of an election." For those who question the accuracy of polls or indeed the reliability of the men who take them, the implications of this assessment are truly alarming.

The pollsters have come to wield such power in large part because many politicians have an almost mystical faith in polling. Richard Nixon, for example, was an early believer in the polls, a fact which was conclusively demonstrated by a bizarre incident during his 1962 campaign for the governorship of California. Nixon was regarded by most commentators as the clear favorite over the incumbent, Edmund "Pat" Brown.

Lou Harris was doing Brown's polling, and as late as August of that year his surveys showed Nixon with a commanding lead. Nixon himself was somewhat nervous about his position, however, having had a tough primary fight. He had beaten his opponent, an extreme right-winger, fairly easily, but the opponent had held off making an endorsement for the general election.

Harris submitted a detailed report to Brown in which he stated that the only way the governor could win would be to entice Nixon into attacking him for being soft on Communism. While such an attack would appease the conservative wing of the Republican Party, Harris concluded that it would alienate many more moderate Warren Republicans, who would switch over to Brown. Unless Nixon made such an attack, there seemed little chance for Brown to win.

The atmosphere in Brown's camp was gloomy; it seemed unlikely that Nixon would say or do anything which would rock the boat. Yet suddenly Nixon lashed out, not just against Brown, whom he accused of being associated with subversives, but against Harris as well, whom he called "a Madison Avenue puppeteer." Harris says, "I have the clippings from that campaign, and if you read them day by day, you can see that Nixon incredibly did just what we said he would have to do to blow the election. And he proceeded to blow it exactly as we had projected."

It was many years later that Harris learned just what had happened. William Rogers, the former secretary of state and longtime friend of Nixon, revealed to Harris that his confidential poll had somehow gotten into Nixon's hands. According to Rogers, when Nixon read Harris' strategy advice, he said, "I know Harris got this report to me intentionally, hoping that I'd do the wrong thing, so I'll outsmart him by doing just what he says will cost me the election."

To this day Harris has no idea how Nixon got hold of his survey. He is just thankful that Nixon outsmarted himself with his Br'er Rabbit logic.

Any other politician who had seen Harris' report would probably have taken it at face value, but Nixon had particular reason to suspect that public opinion polls are not necessarily what they seem to be. During that same California campaign, Nixon and a loyal aide, H. R. Haldeman, personally approved the use of a fraudulent public opinion poll in a deceptive scheme to raise money from Democrats.

A half-million postcards were mailed to Democrats throughout the state by a Nixon front called the Committee for the Preservation of the Democratic Party in California. The cards described Governor Brown as the prisoner of the "ultra-liberal" California Democratic Council and expressed revulsion at that group's alleged support of a nuclear test ban and the admission of Communist China to the United Nations, among other things. People were asked both to make contributions and to express their opinions on a detachable stub. Press releases were later issued which said, "Nine out of ten regular Democrats flatly reject the 'ultra-liberal' California Democratic Council."

The Democratic State Committee went to court to stop the mailings and succeeded in getting a preliminary injunction. Final resolution of the case was not reached until two years later, but when it came, the California court made findings setting forth Nixon's role in the whole episode.

The court ruled, "Mr. Nixon and Mr. Haldeman approved the plan and project as described above and agreed that the Nixon campaign committee would finance the project." Although contributors thought they were giving money to a Democratic organization, the judge found "in truth and fact, such funds were solicited for the use, benefit, and furtherance of the candidacy of Richard M. Nixon for governor of California." Nixon himself, the court said, had prepared the final draft of the phony poll! The incident was a revealing hint of dirty tricks to come, but unfortunately it went unnoticed in the national press.

Nixon had bad luck with polls in 1962, both those he concocted and those he obtained from his opposition. In other campaigns, he was an enthusiastic user of legitimate surveys. In 1968 elaborate polls were conducted around the clock in order to shape the selling of the president. Less than two days after Lyndon Johnson made his dramatic late-October announcement that he had ordered a halt to all bombing of North Vietnam, Nixon had an in-depth analysis of the electorate's reaction.

In 1972 the Committee to Re-elect the President hired Bob Teeter to oversee the most extensive polling operation in the history of campaigning. Teeter's own firm, Market Opinion Research, did much of the surveying, but supplemental work was carried out by Opinion Research Corporation of Princeton, New Jersey, and a California firm, Decision Making Information. Waves of polls were taken in target states, and national surveys were conducted to test the impact of advertising. There was the capacity to get an almost instantaneous reading of the public pulse on any issue or development, but the fact that Nixon ran away with the election from the start meant that the system's full potential was never really tested.

Polling was also conducted on a wide scale for Nixon's 1960 presidential campaign. Claude Robinson, the founder of Opinion Research and a longtime associate of George Gallup, was, in Nixon's words, one of eight "key men" who directed the campaign. Nixon later wrote that Robinson's surveys were "almost miraculously accurate."

The 1960 campaign really made Nixon a believer in political polls. It also provided some lessons for his pollster. After the race was over, Claude Robinson said, "This election showed that there is no essential difference between the merchandising of politics and the merchandising of products." As an apparent afterthought he added, "You can't build a synthetic candidate the way you can a product." The next decade, however, was to bring undreamed-of advances in the technology of synthetic politics.

Public opinion polls are a way of test-marketing a candidate. Sometimes polls are used to guide changes in the candidate which are merely cosmetic. When the polls showed former California Assemblyman Jesse Unruh that his obesity gave him the image of a political boss, he lost ninety pounds and kept it off. Former Presidential Press Secretary Pierre Salinger, running for the Senate from California, was told by his pollsters to stop smoking cigars in public and to reminisce more about his boyhood days in the state in order to counteract his image as a carpetbagger. They added that though these gestures would help, he would still lose, and he did.

A strange crisis over socks developed in Milton Shapp's first campaign for the Pennsylvania governorship in 1966. A survey showed that he was being hurt by his habit of wearing maroon socks. Somehow the report leaked out and was given some publicity. Shapp was left with the choice of either continuing to offend those people who did not like the color of his socks, or getting new ones, thus making it seem that he was the puppet of some anonymous public relations man.

Elections are not often won or lost over cigars and socks. Where there are real issues at stake and the candidates are addressing them, we need not worry about superficialities. What is chilling, however, is the prospect of candidates who say and do only what their pollsters tell them to.

In the 1960's a West Coast public relations firm advertised, "You can be elected state senator; leading public relations firm with top-flight campaign experience wants a senator candidate."

One of their clients, with no previous political experience, ran entirely on the basis of his last name, Campbell. It was not a locally prominent name; rather, the advertising firm exploited the homey associations everyone had with the famous soup. Billboards were produced with the name *Campbell* written out in the same smooth script which appears on the soup cans. There was no speech-making or hand-shaking; the whole campaign rested on the advertising. The gimmick worked, Mr. Campbell was elected, and a bright new political future opened up for the Kelloggs, Marlboros, and Fords of America.

Hal Evry, who runs the firm, does not care about his clients' politics. He requires only that they are rich enough to foot the campaign bills and smart enough to follow his directions. Evry makes it a strict rule that any client of his avoid the risks of personal appearances. "He makes a speech and then exposes himself to foolish questions from some nut who makes him look bad." The idea of exposing a candidate to the people offends Evry and some others in his trade just as much as the thought of operating without first scrubbing would shock a surgeon: the sterility of the enterprise cannot be jeopardized. Evry says that 93 percent of his clients have been elected, but some California politicians are skeptical about his claims.

Evry has "great admiration for Dr. Gallup" and sees public opinion polls as the salvation of modern democracy. He looks forward to the day when selected voters will be polled as to the types of candidates they want "even on physical and personal characteristics." The candidates who most closely fit the profile of the survey will run for office. We need not even bother with a final election: Evry thinks voting is "ridiculous." He says that a lot of time and effort would be saved if the government sampled the nation to find a thousand typical voters and let them decide. "Throw it in a computer, and whatever comes out, that's it."

Perhaps Evry's vision goes a step beyond that of most of the modern political consultants, but even in its more modest forms this politics of image-manipulation has sinister implications. It plays upon the subliminal mind and subverts the intellect. Image-making is dedicated to the exploitation of what could be called the Gresham's Law of Politics: The glib, the evasive, the dogmatic, and the slick will always prevail over the serious, the sincere, and the judicious in any political discussion. The emphasis is on emotions, not ideas or principles.

It is a mistake to believe this kind of politics is limited to Los Angeles or Madison Avenue. Its appeal is overwhelming to any politician who wants to retain his office. Tip O'Neill, the House majority leader, climbed his way up in Democratic politics from the working-class neighborhoods of Cambridge. He is as earthy and straightforward a man as you are likely to find in Congress, but he is quick to admit that if he had a serious challenge, the first thing he would do is hire a professional consultant. "I'd bring in one of the best in the nation to package a Tip O'Neill. I'm not going to change in philosophy. I wouldn't let them picture me as I'm not. But they'd glamorize me, there's no question about that."

O'Neill has not faced a tough contest in years, and when he talks about being "packaged" he sounds a little wistful. A man who feels that he is not fully appreciated by his constituents, he admits to being hurt when he saw a poll which showed that fewer than half of the people in his district knew he was their congressman. The idea of snappy television ads, handsome leaflets and posters, all the trappings of a professional campaign which would make a more glamorous Tip understandably appeals to O'Neill. Campaign consultants exploit this kind of vanity. Other politicians, less sure of who they are, are all the more prone to turn themselves over to some consultant who will tell them what to say and how to say it.

Political packaging and polling go hand in hand, for if the candidate is a product, then it is necessary to know what the voters are buying. People like Joe Napolitan and John Martilla will contract to handle an entire campaign—advertising, canvassing, speech-writing, the works—for a given price. Public opinion polls are part of the deal, for the managers look to the polls to determine how to run the campaign. Joe McGinnis' account of the Richard Nixon campaign in 1968, *The Selling of the President*, sets out the process in detail.

There is only the haziest of lines separating appearance from substance in politics, and it is increasingly apparent that the influence of polls has extended beyond the manipulation of images to the determination of policy.

It is not easy for a politician who owes his existence to winning votes to take a tack contrary to the polls. Hubert Humphrey admits that he is impressed by surveys on issues, particularly where there is a trend one way or another. "Take foreign aid. There was a time when it had sixty-percent support, or even better than that. Now it's down to maybe twenty. So every politician knows that if you vote for it, you're doing it out of deep conviction, not with any political benefit, to the contrary with a political deficit."

Few politicians will admit they actually make policy decisions according to what appears to be popular in the polls, but they certainly do. Lloyd Free, who polled for Nelson Rockefeller when he was governor of New York, has said that while Rockefeller was never "unduly influenced by the polls," he would heed them "when the public speaks loud and clear."

Free specifically recalled that when the issue of public versus private ownership of nuclear power plants was hotly debated in New York in the mid-sixties, Rockefeller originally supported private ownership. According to Free, however, "Our poll showed that the public was overwhelmingly in favor of joint development by private investors and a public authority. On the basis of the poll and other factors—Senator Kennedy came out in favor of public ownership—the governor re-examined his position and finally favored joint development."

Was Rockefeller's shift a responsible concession to the will of the people, or was it a self-serving political ploy? In either event, to the extent that it depended on Free's surveys, it rested on the accuracy of the information Rockefeller was getting from his pollster.

Most pollsters see nothing terribly wrong in a politician's modifying his public point of view in response to what the polls tell him is popular. The late Oliver Quayle was quite candid about what his clients would do. "Sure, let's be honest. They will concentrate on those issues that help them the most. Suppose you are a senatorial candidate who believes in expanded Medicare and a pullout from Vietnam. The polls show your stand on Medicare will get you votes, but Vietnam might hurt you if you pushed it too much. So you concentrate on Medicare. What's wrong with that?"

What is wrong is that it assumes, first, that polls accurately reflect meaningful public opinion, and, second, that politicians are obliged only to echo what they hear as the public will. No doubt Quayle was right: many politicians tell the people what they want to hear, and nothing more. Though politicians may rationalize such conduct as pragmatic, it borders on outright duplicity. If a candidate has a controversial position on Vietnam or some other issue, his electorate has the right to know it.

Pollsters point out that even if there were no polls, many politicians would shy away from controversial issues. They are undoubtedly right, yet the existence of the nationally syndicated surveys and the tremendous growth of private political polling has greatly exacerbated the problem. It is one thing to have an intuitive sense that foreign aid is less popular than it once was, but it is another to confront the stark statistics which seem to indicate with decimal-point precision how dangerous an issue it could be.

Polls on issues are not, after all, very reliable. Even leaving aside the question of whether elected officials ought to do what is popular as opposed to what they themselves think is right, there is no guarantee that the polls can really tell them what is popular. Yet the polls tend to inhibit candidates and officials from breaking new ground and educating the public, and as a result leadership is diminished.

Leo Bogart, the outspoken former president of the American Association for Public Opinion Research, has warned that politicians who try to court public opinion may be chasing a will-o'-the-wisp. "A public official who does something that is disapproved of by a plurality of respondents in a survey risks the accusation that he has defied public opinion, even though the very act of his leadership is likely to shift opinion toward the course of opinion he asserts."

Political polls have become such a dominant feature of modern campaigning that they influence the candidates whether they like it or not. Candidates must practice the art of pollsmanship whether they are way ahead or far behind. For example, when George McGovern appeared on ABC's "Issues and Answers" in March 1972, before the primaries had begun, the first third of the questioning was directed not at his views on the war in Vietnam or his proposals for the economy, but rather to his poor standings in public opinion polls. Like a coach being interviewed at halftime while his team is trailing twenty-eight to three, McGovern was invariably asked to explain why he was doing so badly. When he would persist in seeing signs of hope, he was treated almost like a poor sport for not graciously conceding that his campaign was a lost cause.

Things got even worse in the general election. At times the polls drove all the other political news off the front page. Would Nixon win by a landslide? Would McGovern prove the pollsters wrong again, as he had in the primaries? Was Daniel Yankelovich right in saying that the polls might encourage Democrats to defect to Nixon? All the press seemed inclined to write about was the horse race and the handicappers. Democratic vice-presidential candidate Sargent Shriver, fed up with the repeated questions, snapped at one interviewer, "The important issues are being obscured by people like us talking about the polls."

As Hubert Humphrey observes, however, "Polls are part of the political environment, just as much as falling farm prices or an endorsement from the AFL-CIO." As such, every successful politician must know how to cope with them. Richard Nixon was an adept practioner of the art of pollsmanship, particularly a variant which might be called "poor-mouthing." In 1968 it was clear that Nixon was going to win all the Republican primaries by default. George Romney, who had been considered the front-runner for more than a year, was persuaded to drop out of the race before a single ballot was cast when he saw surveys which showed that Nixon would beat him in New Hampshire by better than five to one. Rockefeller vacillated about running, and when he finally decided he would, he declined to enter the remaining primaries.

Ordinarily this lack of opposition would be welcomed by any politician, but, having lost the presidency in 1960 and the gover-norship of California in 1962, Nixon needed to prove himself a winner; he had not won an election in his own right since 1950. In order to have the semblance of a contest, Nixon would an-nounce in advance the percentage of the vote he hoped to get. His estimates were always coyly modest; he had no organized opposi-tion, just favorite sons and write-in candidates. Nixon's private polls told him that he would get 70 percent of the vote in New Hampshire, but he said with a straight face that he would be satisfied with 50 percent.

This ploy seems transparent, but Nixon got some mileage out of it. That his forecasts were always low publicly demonstrated his humility, not to mention the apparent snowballing of his popularity. By inflating the impact of the final vote, Nixon hoped to obscure the fact that he actually had beaten no one, and to an extent at least it succeeded.

Poor-mouthing can sometimes backfire. Several weeks before the 1968 Democratic primary, New Hampshire Governor John W. King thought he was being safe in saying that he would be "disappointed" if Eugene McCarthy got more than 25 percent of that state's vote; at the time all the polls showed the Minnesota senator getting far less than that. Yet in the last weeks of the campaign McCarthy's efforts caught fire, and he ended up with more than 40 percent of the vote. His performance looked all the better because it had substantially exceeded the estimates made by supporters of President Lyndon Johnson, like King.

Again in New Hampshire, four years later, Maria Carrier, a prominent local Democrat working for Muskie, said that she would "cut her throat" if he did not get at least 50 percent of the vote. When she made that statement Muskie had 65 percent, according to the *Boston Globe*'s poll, but as it turned out, that support was not very deep. Muskie came in with 47 percent, certainly a respectable figure, yet it did not meet the standards which even his own people had set up.

In all this pollsmanship it is often the pollster who ends up getting used. During the 1960 presidential campaign, Lou Harris quite willingly let Kennedy exploit his name for propaganda purposes. In 1972, one of Pat Caddell's principal functions for McGovern was to deal with the press, to try to counteract the bad news coming from other pollsters. Caddell prefers to think of his role as that of educating the press, but in the end, it was his reputation and good will that was being tapped.

During the 1972 primaries Caddell teamed with Frank Mankiewicz to give special morning-after seminars to the press on how to interpret the election returns. In advance of each primary, they would select ten to twenty precincts throughout the state, and prepare elaborate fact sheets on each one, setting forth ethnic compostion and past voting patterns. Then, shortly after the results were in, Mankiewicz would summon reporters to a press conference. He recalls, "I'd say something like, 'Okay, here in Precinct Thirty-four, a working-class district in Kenosha, 54 percent Slavic, McGovern ran strong.' The press ate it up. We never would have persuaded them that we were geting the blue collar vote if we hadn't made these presentations."

The information that Mankiewicz handed out was accurate, but carefully selected. "If McGovern did well in two out of three of the Slavic districts we had pre-selected, you can be damn sure that we didn't tell them about the third." The presence of Caddell, the Harvard-educated pollster, gave these seminars an aura of authencity. Mankiewicz cheerfully admits, "We didn't do it, but we could have made it all up—the ethnic and class make-up, even the results! No one ever checked up on us. We were doing the reporters' work for them."

One of the most blatant examples of a client exploiting a pollster involved Archibald Crossley in 1967. At that time Lyndon Johnson was doing badly in the national polls. The most recent Gallup Poll had him trailing Rockefeller, Romney, and Nixon in trial heats. A Crossley survey, however, was leaked to Drew

Pearson and found its way to the *New York Times* and other papers. It ostensibly showed that Johnson enjoyed a lead over four possible Republican candidates—Rockefeller, Reagan, Romney, and Percy—in a so-called bellwether New Hampshire county and that he trailed Nixon there by only a single point. Stafford County, New Hampshire, was said to be significant politically because in the past twelve presidential elections it had reportedly voted only about one point more Democratic, on the average, than the country as a whole.

At first it was not revealed who had commissioned the survey or how it had been taken. Crossley was rumored to be indignant about the way it was being used, but at the outset he would not divulge who his client was. He stated, "It's supposed to be a confidential relationship, and I have kept my part of the confidence." Crossley eventually yielded to pressure from both the press and many of his fellow pollsters, particularly when it became obvious to him that his name was being deliberately used to lend the impression of authenticity to a deceptive report.

When the real story was told, it was clear that, rather than demonstrating Johnson's strength, the poll confirmed his vulnerability. The survey had been sponsored by Arthur Krim, a New York Democrat who was very close to the president. Krim had specifically instructed Crossley where he was to poll, and he could hardly have selected a more hospitable place to search for whatever was left of the consensus which had elected Lyndon Johnson three years before. In 1964 Johnson had received about 61 percent of the vote nationally, while in Stafford County he had won 68 percent. Thus the fact Johnson that still held an edge there over most of his rivals was absolutely no test of his national appeal. Indeed, that the Republicans ran as well against him as they did made his chances look bleak in the rest of the country.

The Crossley survey had been carefully constructed to produce results favorable to Johnson, and only part of the information, at that, had been leaked to the press. Crossley later said that he did not think that Krim had been responsible for the misleading leaks, but that they had come from the White House itself.

The only defense a pollster has against having a client exploit his name is the threat that if the client releases part of the poll in a misleading manner, the pollster can release it all. Some pollsters specifically provide for this in their contracts. In 1974 Richard Wirthlin, president of Decision Making Information, had to drop a candidate during the height of a campaign. Wirthlin had taken a poll that showed his client well behind. The candidate later took his own crude survey, which ostensibly showed the race to be a toss-up. The candidate released the polls, claiming that both of them had been conducted by Wirthlin. Wirthlin demanded a retraction, got it, and still dropped out of the race. "The fellow wanted to use our name to give a distorted impression of his strength, but we know that he didn't have a chance of winning.

Not all pollsters, however, are willing to blow the whistle on their clients. Some have even been known to offer their clients a package of two polls, an accurate one for their own confidential use and a trumped-up one to be leaked to the press. It is a devious but effective way to get business.

When Louis Harris was doing private political polls, he lived by an ethic which provided that loyalty to a client is a pollster's one and only duty. In the 1962 race for the Democratic nomination for governor of New York, for example, Robert Morgenthau's staff selectively leaked parts of a survey Harris had conducted for them. The fragments implied that Morgenthau would be a stronger candidate against Nelson Rockefeller than any other Democrat, but the wording of the actual questions, their sequence, and the sampling methods were not revealed. Because of Harris' recent associations with President John Kennedy and New York Mayor Robert Wagner, his work carried a great deal of weight within the Democratic Party.

Morgenthau's opponents, among them James Farley, questioned the validity of such scanty information, and Harris came under heavy criticism for not disclosing the entire poll on his own, but he refused to buckle. Morgenthau, it turned out, was on the board of directors of Harris' firm! Morgenthau eventually won the nomination, but Harris had alienated many people, both pollsters and politicians. Elmo Roper shortly thereafter noted "a suspicion that the polls are in some cases being *purposely* and *primarily* used to influence the voting outcome. I am afraid that this suspicion is not altogether unfounded."

Harris, it should be noted, contends that press accounts of the episode were deliberately slanted to make him look bad. "I had gotten very powerful in New York State because of the work I did for Bob Wagner in 1961 when nobody gave him a prayer of winning. A whole bunch of machine types whom we had ousted were out to get me. They had access to part of the New York press. It is as simple as that."

Pollsters are not entirely to blame for their own exploitation. Political polling involves pressures which are hard to resist. Elmo Roper's son, Burns Roper, says that his firm now tries to avoid political polling because clients so often want to distort the data. "If you come up with three different ways of measuring candidate strength, and the client looks lousy on two of them but looks good on the third, he wants to ignore the first two and just publicize the third one, saying, 'A Roper study shows me out ahead.' "

Roper thinks that it's a no-win situation. You gain no friends by releasing all the data and making your client look bad, but if you keep quiet, your reputation as a pollster can suffer if the client ends up falling on his face. Roper concludes: "Frequently their purpose is just publicity. They don't give a damn if the data is any good; they just care if the results are to their liking. And they want to be able to use some pollster's name that has acceptance in political circles."

It is all too easy for a political pollster to lose his perspective in a campaign and identify his interests with those of his client. Pollsters usually have a direct line to their clients, be they first-time candidates for Congress or presidential nominees. Indeed, during the heat of a campaign their clients are often after them. In the uncertainty of a long and hard battle they seek out any kind of word on how they are doing. Just days before an election, a candidate for the Senate asked the late Oliver Quayle to conduct a last-minute poll. Quayle counseled against it, noting that it was really too late to make any changes in strategy. The candidate was insistent, however, pleading, "I just have to know."

The pollster supposedly possesses valuable information. He is stroked by his client and courted by others of the staff, as well as those in the press. It is easy for the pollster to convince himself that he is indispensable and that his client's fortunes ride on his talents, even, perhaps, that history is at stake.

Peter Hart, a young Washington pollster specializing in political work, admits that it is a heady business, one that can easily give you a big ego. Hart, however, learned his lesson early. When he was twenty-seven years old and working for the Democratic National Committee, he was invited to lunch by columnist Joseph Kraft. They went to Sans Souci, and Kraft ordered entirely in French. Hart was practically giddy with delight at the thought that Kraft wanted to pick his brain. While they were walking back to Hart's office, Kraft glanced into a drugstore window and said, with a little disdain, "Oh, there's George Reedy." Reedy had been Lyndon Johnson's press secretary a couple of years earlier; now he was yesterday's news. Hart said the incident brought him back down to earth then and there. "Pollsters may be important people up to the election, but they're nothing afterwards. If you start believing otherwise, it's kind of sad."

Perhaps we should not worry too much about what politics does to pollsters. They are, after all, consenting adults. But we should be very concerned about what pollsters do to politics. The trend toward packaged politics based on polls produces homogenized candidates with warm smiles, steely yet sincere eyes, and tongues which never speak to the hard issues confronting the nation.

The notion that deference to the polls is somehow in keeping with democratic government is a dangerous one. Elmo Roper, unlike many of his colleagues, cautioned against making policy on the basis of polls, for polls often do not represent a mandate from the people. "Majority opinion is often based on inadequate facts and sometimes on impulsive emotions."

Leo Bogart, quoted earlier, has warned that principles which may be appropriate in business are irresponsible in politics. "In the domain of consumer research, it is easy enough to say that a manufacturer should give his toothpaste the flavor preferred by a majority of his customers. But the tendency to regard policy as a commodity that should obey the laws of supply and demand becomes scandalous when it is extended to the realm of politics."

To say that politicians should not make policy by the polls is not to say, as some pollsters imply, that public opinion must then be irrelevant to decision-making. Broad moods of concern may originate with the people, as was the case with the war in Vietnam and environmental problems at home, but it is ultimately the task of political leadership to articulate these moods and to define alternative policies in response to them.

A second danger of political polls is their effect on candidacies. Superficial surveys can kill campaigns before they start or overinflate the influence of people who have the good fortune of being well known. There are politicians who have licked adverse polls, but they are rare. In 1972, when Dick Clark of Iowa began his race for a Senate seat, he was known by so few people that he did not even bother to take a poll to test his visibility. He won an upset victory against an incumbent Republican by running a vigorous campaign which included an attention-getting walk across the state.

Dick Clark's experience, however, is the exception which proves the rule. Clark had been an administrative assistant for Congressman John Culver. Culver himself had planned on running, but adverse polls helped convince him not to. Only then did Clark enter the race. Perhaps taking heart from his protégé's success, Culver ran for the other Senate seat in 1974, and he too won.

For every Dick Clark there are likely countless others who are never able to get off the ground. People who might have provided wise and compassionate leadership have never come to the fore because negative polls prevented them from getting the financial support and personal assistance necessary to mount a campaign. If there is any consolation, it must be in the thought that the polls have also killed the hopes of some scoundrels and thieves.

We can never know just what we have missed. Tip O'Neill once seriously considered running for governor of Massachusetts but decided against it after seeing polls which showed that he was not known statewide. In 1962 Pat Brown almost decided not to run for re-election as governor of California when he saw a Harris Poll which showed him trailing Nixon by a wide margin. Some leaders of the anti-war movement in 1968 tried to talk Eugene McCarthy out of challenging Lyndon Johnson in the Democratic primaries on the ground that McCarthy was such an unknown he would surely do badly, and that would reflect poorly on the peace movement. For every Brown and McCarthy who do take the plunge and do well, there are probably many more O'Neills who do not even try.

The reliance on surveys to test the legitimacy of candidates pervades the political world and beguiles even the people one would expect to be most skeptical. Before the 1972 Democratic convention George McGovern studied polls conducted by Pat Caddell testing who would be the strongest running-mate. What irony! Here was McGovern, who himself only a few months before had been rated in the single figures by the national pollsters, looking to polls to determine which candidate would help the ticket the most.

McGovern claims that he did not tell Caddell to take the surveys, that Caddell undertook the project on his own. Nevertheless, once the results were submitted, the McGovern campaign staff studied them carefully. Indeed, they were so fixated by the polls that they ultimately made what proved to be a disastrous choice.

In essence, Caddell's surveys said that, with one exception, it did not matter who the vice-presidential nominee was. McGovern ran about the same, whether it was with Thomas Eagleton, Edmund Muskie, Pat Lucey, or any of the other often-mentioned possibilities. The one exception, of course, was Edward Kennedy. With Kennedy as a running mate, McGovern ran much more strongly, both nationally and in key states like California.

The Caddell poll created a mentality among the McGovern people that they simply had to have Kennedy. To the very end they held out hope that he could be induced to come aboard, and some of them were surprised when he did not. They had shown him the survey results to try to convince him that he had a duty to his party to run. Kennedy says now, with a trace of contempt, "I saw Caddell's polls, but there was no way I was going to get involved in any of that."

Largely because McGovern's staff counted on getting Kennedy, it failed to do its homework on other possible candidates, a failure which proved critical after the hasty choice of Senator Thomas Eagleton.

McGovern was not the first nominee to consider the polls in picking a running mate. In 1960 Richard Nixon had his pollster,

Claude Robinson, test a number of possible combinations. Robinson found that there was not much difference between Henry Cabot Lodge, Thruston Morton, and other possibilities, again with one exception—Nelson Rockefeller. Nixon made a clumsy attempt to get Rockefeller to run with him, but Rockefeller declined, later announcing publicly that he was not the type to be "standby equipment."

Political polling has grown and changed radically in the past two decades. Mark Hanna used crude polls before the turn of the century, and Senator Jacob Javits used polls when he first ran for Congress in 1946. An editorial in the *New York Herald Tribune,* in fact, congratulated Javits on what it regarded as a great step forward: "This represents a considerable advance over the old torchlight processions, the street-corner tub-thumping and rabble-rousing, which used to constitute the major weapons in political campaigning." It was really Lou Harris' well-known studies for Kennedy and Claude Robinson's less visible but equally important work for Nixon in 1960, however, which marked the advent of modern political polling.

In 1960 few candidates for Congress invested much in professional polling, but in 1966, 85 percent of winning senatorial candidates had depended on polls. In 1968 it was conservatively estimated that political candidates spent four to six million dollars, perhaps more, just on polls. According to Pat Caddell, the most expensive elections were those held in 1970 and 1972. There was a down-turn in spending in 1974, he says, in large part because the Watergate scandal had made people much more cautious about both giving and receiving political donations.

The future of political polling is not entirely clear. Business has remained good on the Republican side—both Ford and Reagan were polling intensively more than a year before the 1976 presidential election—but many Democratic pollsters have been suffering. Going into the election year, Caddell, Peter Hart, and Bill Hamilton were all waiting to be retained. They unanimously attributed the drop-off in business to the fact that most of the candidates seemed underfinanced. At the time, Hart said, a bit wistfully, "You know, polling could be a heck of a planning vehicle when nobody else has any information on who the target groups are and what they need to appeal to them."

In theory, tight money might be expected to cause even greater reliance on polls, as it is essential to know how to allocate limited resources most effectively, but the pollsters had relatively little success with this pitch. As Caddell notes, "A dollar spent on polling is a dollar taken away from communications."

The slow start in 1976 probably does not represent a trend, however, for the political situation at that point was rather unusual. Federal legislation had restricted both contributions and expenditures until January, when the Supreme Court struck down much of the law. Prior to the court's decision, Caddell had feared that the law's impact would be the greatest in congressional races, where the spending limits had been particularly severe. Had the law stood, candidates would have had to pool their survey data or resort to cheaper methods, such as telephoning and using volunteers for interviewing. Given the great number of congressional races, the federal limitations could have had a more devastating impact on pollsters on this level than in presidential campaigning.

The bleak financial picture had been compounded by the fact that there were so many contenders in the race. Although the total amount contributed was larger than most observers realized, it had to be split among many different candidates. In the future, with a narrower field and fewer financial restrictions, the political survey business will probably rebound.

Even if political polling does enter a period of retrenchment, it is likely that the Republican pollsters will feel the pinch less than their Democratic counterparts. Bob Teeter, who has polled for Ford, Nixon, and a host of other Republicans, says: "Republicans traditionally have been more advanced technically in campaigning; they feel a greater need for this kind of information. Over the long run, spending limits may keep polling from expanding, but we never really have taken that large a share of a campaign budget."

10

The Nielsen 1200

Television news directors and politicians are involved in an incestuous relationship. The pollsters did the pimping.

TV news, once a stepchild of local programming, is now a station's champion in the continual battle for higher ratings. The evening news supposedly attracts viewers who, once having tuned in, stick with the same channel throughout the night.

The stakes are high in this battle to top the ratings. More than 95 percent of all year-round household units have at least one television; the vast majority have more. According to Gallup, television, not baseball, is America's favorite pastime. Studies by Opinion Research Corporation indicate that television is the most trusted news source. If you like and trust your newsman, so goes the argument, you will like and trust his station, and by implication, the products advertised on it.

Competition for viewers has all the subtlety and good taste of a gas war between abutting service stations. Anchormen become stars, working their way up to bigger and bigger parts. After apprenticing for several years in Boston, Tom Ellis is tapped to go to New York at a salary greater than that of the president of the United States. There, Ellis has the formidable task of going head to head with Channel 2's Jim Jensen. Jensen, in turn, is touted by his own station as "New York's longest-running anchorman," as if he were a Broadway institution, like *Hello, Dolly.*

The news itself is shaped and packaged to make it more popu-

lar. Happy news has replaced journalism. Banter between the
anchorman and the weatherman replaces analysis and interpretation. Special guest commentators are hired not for their insight
or intelligence but for their entertainment value. The end result
is a fast-paced, upbeat hour of gossip and quips, not much different from the style of "Laugh-In" of several years back. Why all
this frenzy? Because it produces the best ratings.

TV news directors live or die according to the ratings their
shows draw. No matter how high the quality of the production
is, if the ratings are low, the director will be let go. If television
ratings are the currency in which their value is measured, it is
not surprising that news directors have unquestioning faith in all
public opinion polls.

Political polls are treated very carelessly on television, even
more so than in newspapers. Walter Cronkite will typically report: "President Ford's popularity has gone up since he explained his pardon of his predecessor, Richard Nixon. The
Gallup Poll reported today that Ford's approval is up three
points." On television, the polls are regarded as being as precise
as the price of wheat or the death toll on Memorial Day weekend.
There is never any caveat that a three-point shift of that sort is
well within the bounds of normal sampling error, nor is there
ever any suggestion that approval polls are not a very good measure of a president's real support.

Similarly, within days of Patty Hearst's arrest in September
1975, John Chancellor of NBC reported the results of a special poll
in which people were asked whether she should be released on
bail. No mention was made of the actual wording of the question,
nor was it revealed whether people who were not familiar with
the case or did not understand the principles by which bail is
determined were first weeded out of the sample.

Television is a very bad medium for the reporting of polls
because it demands emphasis on the dramatic and the visually
interesting at the expense of things which are subtle and ambiguous. For a poll to get on the air, it must be timely and newsworthy. It need not be correct.

Many local stations promote an evening "telephone poll" or
"question of the night" in which viewers can record their yeas
or nays on one proposition or another by phoning in their re-

sponses. The results are usually announced later in the program or on the next newscast. One night the question might involve gun control, the next divorce laws. The final percentages in any case are meaningless since special-interest groups can flood the switchboards and completely distort the results. In spite of their inaccuracy, such surveys are considered a good feature because they stimulate viewer involvement.

The typical attitude of newscasters toward reporting public opinion polls is summed up in a remark by a television pollster, quoted by the former president of the American Association for Public Opinion Research, Leo Bogart: "We know the figures are wrong, but we want to give them to you anyway."

Even though the networks give little air time to public opinion polls and are careless with the ones they do present, television spends a great deal of money on surveys. The networks have their own research departments and frequently hire big-name pollsters to handle special projects. Almost none of the data which is gathered is ever reported to the public, yet it has a profound effect on what stories are covered and how they are treated.

The Super Bowl of television news reporting occurs with national elections. Each of the networks pulls out all the stops in order to win the ratings battle. It is estimated that in 1974, for an off-year election, the networks spent eight or nine million dollars just to produce the election-night coverage. In presidential years, the amount is correspondingly bigger.

In the past decade the contest among the various networks as to which will be the first to project the winners has almost overshadowed the elections themselves. Commentators drone on about turnout and ethnic voting patterns, but their function is merely to kill time while the computer digests the early returns and predicts the winner. The network which comes up with a projection first has scooped the other two. Its commentators can pontificate on the meaning of the results, while their rivals are still speculating about what will happen.

Each network uses somewhat different methods to make their projections; in fact, the methods often are changed from election to election. The basic theory, however, is constant. The pattern of election returns as they trickle in is not necessarily representative of the total vote. Citizens in villages like Hart's Location, New Hampshire, cast their ballots right after midnight, so that for a few hours, at least, their votes dominate the national totals. Obviously the fact that 80 percent of the populace favors one candidate or the other is meaningless when there are only eighteen voters in the town.

Even when half the votes are in, the outcome may still be in doubt. Returns do not come in randomly. Usually the urban areas, with voting machines, report first while the tallies from areas which still use paper ballots may not be available for hours. A candidate may carry the cities and have an apparent lead several hours after the polls have closed, only to see it melt away as the votes from slower-reporting rural areas are tabulated. Thus the networks cannot simply look at the first 1 percent of the vote that comes in and assume that it is representative of the rest.

The early returns are nevertheless the basis of the projections. The networks correct for the fact that they are not representative in one of two ways. Some have used conventional polling techniques, randomly selecting in advance a number of precincts in a given state. In a statewide election perhaps 2 or 3 percent of the precincts would be selected; in a national contest, 3,000 out of the total 108,000 would be used. In essence, the selection process is identical to the one that Harris and Gallup employ to select geographic areas in which they will poll. Chance sampling error is subject to the same control; indeed the networks usually sample enough precincts to make the possible margin of error substantially less than that of the newspaper surveys.

There is, however, a practical drawback in conducting a poll of this sort if the overriding objective is to get the results fast. The networks do hire local people, at modest pay, to telephone in the results as soon as they are reported, but some areas are bound to report later than others and this will hold up the projection.

To deal with the problem, some networks have used a second approach. Instead of selecting sample precincts at random, the analysts choose them according to their characteristics, the most important of which is whether they tend to report early. Even if the selected precincts do not vote exactly as the entire state does, they may exhibit a consistent pattern, such as being four to six points more Democratic than the statewide vote in off-year elections. If that pattern is repeated, the precincts may be just as good an indicator as a watch which you know is always five minutes fast.

The problem with this method, however, is knowing whether the past is prologue. Suburban Boston used to be heavily Republican, but in recent years there has been a liberal trend in communities like Newton and Wellesley. In the past, if a Democrat held his own in such areas, it was safe to assume that he carried the state, but now the pattern has changed. There is, then, a degree of subjective judgment involved in the selection of the sample precincts and the weighting of them.

Because of the pressure to get information in, some of the local poll-watchers are bound to make mistakes. An erroneous report from one or more precincts could throw off the projection markedly. In order to guard against this possibility, some election analysts construct their computer program so that any results which seem implausible are knocked out of the model.

This safeguard, however, can backfire. In 1966, for example, CBS projected that the Democratic candidate, George P. Mahoney, would beat Republican Spiro Agnew for the Maryland governorship; their computer had systematically rejected all the reports of black precincts going for the Republican, Agnew. Now they have developed more sophisticated programs which ostensibly can assimilate unexpected trends, but there is still as much personal judgment as hard science in making projections, and mistakes are inevitable.

Even though the projections are not infallible, they have come to dominate election-night coverage. All the reporting revolves around the computer's forecasts. In one moment the newscasters are describing the start of an exciting contest; in the next they are doing a post-mortem. Though some politicians persist in holding out hope after the networks have pronounced their candidacies dead, more and more are meekly accepting the judgment.

In 1974, for example, the networks projected that John Gilligan would be re-elected governor of Ohio. In spite of the fact that the race had been bitterly fought, his Republican opponent, James Rhodes, gave a long and gracious concession speech in which he vowed to cooperate with Gilligan for the betterment of the state.

Rhodes accepted the condolences of friends and supporters; he tried to hold his head high in defeat. But after midnight, things started to change. As the returns piled up, the race tightened. As the night wore on the shift continued, and Rhodes, in spite of his concession, was elected governor!

If the election projections merely caused candidates to make premature concessions, then there would be no reason to be concerned with them. They would be just another kind of forecast, like the weatherman's, to poke fun at when it goes wrong. Yet unlike the weather forecast, which has no effect whatsoever on tomorrow's weather, the projections could influence the outcome of presidential elections.

Now that rapid reporting from the sample precincts and instantaneous calculation by computer allows the networks to declare a winner within minutes of the closing of the polls, it is possible to have projections for the eastern states hours before the polls close in the west. In a close election, where the electoral votes of California or other Pacific states could throw the election one way or the other, the impact of these television projections could be decisive. People trying to decide whether to vote at the last minute might well be dissuaded if projections made it appear that the election was already over. Perhaps others would be spurred into voting for the apparent winner.

For the most part the networks have denied that their projections could have any effect on elections, arguing that such an effect has never been conclusively proven. This is not, however, a very satisfactory position. In essence, it says, "Show us a presidential election which we have conclusively screwed up; then we will change our ways." It is entirely possible that up until now the effect has been negligible. Perhaps those voters who hopped on the apparent leader's bandwagon were matched by those who went for the underdog. But can we count on things neatly balancing out in the future? If politicians give projections enough credence to concede defeat, might not some potential voters also be swayed by them?

From time to time one or another of the networks have made statements to the effect that they plan to play down the projections and drop out of the race to be the first to call the election. But once election night arrives, the race is back on again, and woe to the network which finishes second. It seems unlikely, then, that the networks will voluntarily cease calling the outcome of eastern races before the western polls have closed. An outright ban on such forecasts might be unconstitutional. The only answer would seem to lie in federal legislation prescribing a uniform moment for the polls to close, such as eleven o'clock eastern time and eight o'clock in the Pacific zone. If balloting ceased simultaneously, there would be no reporting of returns until everyone had voted.

Richard Salant, president of CBS News, has proposed a twenty-four-hour voting day with uniform opening and closing. Among other things, he asserts that this would "eliminate once and for all the national suspicion that publicizing the vote returns where the polls have closed might influence voters in areas where the polls have not yet closed." The mere possibility of such an influence warrants action to prevent its occurrence.

If the impact of projections on voting behavior is still speculative, the impact on election coverage is clear-cut, though oddly paradoxical. The computer projections were originally introduced to jazz up the presentation, to give it a space-age flair. The end result, however, has been just the opposite. The projections have taken all the drama out of watching the returns as they come in, first in an erratic trickle, then in an increasing rush. The sense of struggle is lost. It is almost as if the winner of a fight were declared at the weigh-in.

The pressure is on each network to make earlier and earlier calls. In 1974 Walter Cronkite announced that the CBS computer had declared Arkansas Democrat Dale Bumpers the winner in his senatorial fight, but then sheepishly admitted that the network had absolutely no voting totals to report. The earlier the networks make the calls, however, the sooner the show is over.

Though election-night coverage is fast making itself unnecessary, television is an increasingly important part of political campaigning. Sophisticated campaign consultants know the nature and needs of television news and have learned to exploit them. Press releases and interviews are passé. A clever campaigner must undertake the kind of activity which makes good television news.

The ratings have come to determine what is news and what is not. Presidents once were able to get coverage on all three networks simply by giving a speech or holding a press conference, but Gerald Ford has discovered that the networks are increasingly reluctant to abandon or even delay their regular programming for such events. In October 1975 both CBS and NBC turned down Ford's request for time to give a speech on tax policy. ABC telecast the speech but got a rating of only 8.4 percent, compared to 29.4 percent for "Rhoda." Several days later ABC declined to cover a presidential press conference which conflicted with its highly rated "Barney Miller."

More and more local stations are now buying expensive mobile camera equipment, "creepie-peepies," which greatly increase their ability to present on-the-spot news. Live coverage during the regular news show is more dramatic than taped material. A politician who arranges his schedule so that he is doing something "filmic" at 6:15 will now have a much better chance of being covered on the local news. It is in his interest—and the station's —for him to stage a media event.

Instead of giving a speech on the need to establish a two-hundred-mile territorial limit to protect the fisheries, the candidate should go down to the docks in the evening as the boats are unloading their meager catch. What is good for television news is good for the candidate, but it is not necessarily good for the electorate. Pictures become more important than words. Appearances overwhelm issues.

In 1960, when television was young and poorly understood by most politicians, John Kennedy realized how it could make trivial gestures seem important. He said, "This business of going to plant gates at five o'clock in the morning is a lot of bunk. I could expose myself to twice as many people in other places. But it gives me the appearance of activity, of being up at five o'clock, hustling for votes."

Joe McGinnis' *Selling of the President* documents the extent to which campaigns can be built on television, particularly paid advertising. The success which Richard Nixon enjoyed in 1968 was not lost on other candidates; indeed techniques have been much more sophisticated since then. Campaign consultants and time-buyers work hard to squeeze the most exposure from every media dollar spent. Roger Ailes, who masterminded Richard

Nixon's use of television in 1968 says, "No candidate can win an election anymore without a media consultant." Certainly few candidates seem willing to try to prove him wrong.

An effective consultant tries to make sure that his candidate's advertising is consistent with the rest of his public image. Robert Squire, who was director of communications for Edmund Muskie in 1972, had a client who had been portrayed in the television spots as being calm and judicious, but who became known publicly for being quick-tempered and emotional. Squire looked enviously at the television work which was being done for George McGovern. "There is a unity to the McGovern stuff— sometimes you have trouble telling his news from his spots. It's the ideal situation. John Lindsay's problem was that his commercials were better than his news. The media people were getting too far out in front of the candidate."

The necessity to dovetail paid and unpaid television will likely become even more important in the future. To the extent that federal and state campaign regulations have been left standing by the courts, the amounts spent on advertising may be significantly reduced. A candidate who wants additional television exposure will have to get it by doing something which helps the newscaster's television ratings. It is a neat circle held together by the polls.

In 1970 Nixon's media man, Roger Ailes, orchestrated the New York senatorial campaign of James Buckley. Ailes believes that an election-eve television appeal was the key to Buckley's victory. "The ratings the next day told us he had reached five hundred thousand people with that show." In a telling slip of the tongue, Ailes continues, "And he won that election by a hundred and forty thousand viewers."

Though the battle for the news audience has been especially intense in recent years, the ratings are equally critical for all other kinds of television programming. The television ratings determine what programs we get to see, how much advertisers pay for a minute of network time, and, as a consequence, what we consumers pay for the products which are advertised. Indeed, the price of a share of a network's stock often traces the ups and downs of its overall ratings.

Most people are aware that ratings can make or break a show, but few know much more than that. The rating services are really specialized pollsters. Like their counterparts who report on the political world, they have a great influence on our lives, yet precisely what they do remains largely a mystery.

Although there are many organizations which measure the television audience, the name *Nielsen* is synonymous with the word *ratings*. It is as dominant in its field as Coke and Xerox are in theirs. Les Brown, who writes on television for the *New York Times*, has stated: "The Nielsen Company, which produces the television ratings that influence nearly every program decision made by the networks, is not only the scorekeeper of network television, but the score itself. There are other rating services, but only the national Nielsens are considered to be official by the advertising industry. One trusts the score; there is no other way. Nielsen is the gospel."

There are a number of reasons why the Nielsen firm so completely dominates the television rating world. For one, it got an early start. Arthur Nielsen founded his company in the early forties to monitor radio listenership. With the advent of television, he simply borrowed many of his radio techniques and applied them to the new medium. Getting this foothold was certainly a factor in Nielsen's success, but it is not the only explanation, for there were also other companies, like Hooper and Pulse, which had also done radio work, tried television research, but ultimately withdrew.

It is the nature of the business that one firm should dominate. The networks are the principal sponsors of audience research. Each network cannot simply produce its own estimates of its viewing audience, for such data obviously would be suspect. Thus it is in the networks' interest to have a single, ostensibly neutral scorekeeper to report on what programs the American people are watching. Were there more than one authoritative rating organization, there would undoubtedly be conflicting results.

Nielsen got to be that scorekeeper largely because he owned rights to a "little black box" which can mechanically monitor when a television is on and to what channel it is tuned. The device, called an "audimeter," was designed by several MIT professors to measure radio audiences. When attached to a radio, it recorded on a moving tape how the radio was used. The Nielsen firm would then periodically scrutinize the tapes to come up with its ratings.

The audimeter was adapted to television and has gone through a number of refinements. Until recently, it was neither little nor a single unit; in fact, one part of it was as large as an old-fashioned console radio. The present generation is much more compact. A modern audimeter consists of two small units, each the size of a cigar box, which are wired to the television set. In most instances, they are placed out of sight, in the hope that in time the viewer forgets that they are even there.

To produce its national ratings, the Nielsen Company attaches these audimeters to the television sets in twelve hundred selected households throughout the country. It is this shadow audience of twelve hundred homes which electronically determines the fate of every network television program. The homes are chosen in much the same way that people are picked to be interviewed in the Gallup or Harris polls. In theory, every household has an equal chance of being selected.

Each Nielsen household is given twenty-five dollars for joining the sample and receives a few dollars a month, the exact amount depending on the number of sets in the home. Nielsen also picks up half the cost of television repair. The company regards this as nominal compensation for being allowed to enter people's houses. If the amount was any greater, Nielsen fears that people would begin to think they were being paid to watch television, and this would bias the results.

The latest audimeters are very sophisticated. They check the tuning status of the television every thirty seconds and make a record of every time there is a change. The audimeter is tied into the phone lines, and at least twice a day a central computer automatically telephones the audimeter and collects the stored data. The circuits and switches dutifully perform their tasks; there is nothing that the viewer must do to record his preference other than turning on the set.

The Nielsen computer contemplates the bits of information it has gathered from audimeters all across the country and comes up with the ratings. It actually has the capacity to report almost instantaneously what channels the sample sets are tuned to, but

because of time zone differences and other discrepancies in sche-
duling it takes some time to translate this information into a
measure of who is watching what programs. Nevertheless the
computer works fast enough. A new program which premieres
in mid-September may be canceled before October because of
bad ratings.

The computer calculates two related measures of viewership.
The first is the "rating," a statistical estimate of what percentage
of the almost seventy million television homes in the country
watched a given program. According to Nielsen, on a typical
evening approximately two-thirds of the American households
with televisions will be using them. Thus, if everyone in this
audience was watching the same program, it would have a rating
of 66 2/3 percent. If all three networks were to split this audience
equally, they would each get a rating of about twenty-two.

Because some viewers are lost to local programming, UHF
channels, and public television, the networks do not really get to
split the entire pie. Nevertheless, a prime-time program which
pulls a rating of seventeen or less is almost certain to be cancelled
even though it may have a national audience of more than ten
million households and twenty million viewers.

The second measure of audience is the "share." While the
rating is based on all television households, the share is based
only on those households which are watching television. The
share thus provides a better picture of how a given program is
doing against its competition. If a program drops a point in its
share, that point has to go someplace else. The old rule of thumb
was that a program with a share of less than thirty was in trouble,
but now some shows with a point or two less are retained if they
have specific appeal to a desirable audience.

The advantages of the audimeter, and the competitive edge
which it gave to Nielsen, become apparent when other methods
for measuring television audience are considered. One method,
the "telephone coincidental," involves calling people up and asking them what they happen to be watching at that very moment.
Hooper used this technique to a great extent, but it has two major
drawbacks. First, the researcher cannot make calls very late in
the evening without annoying a sizable number of people who

are trying to get their sleep; hence the telephone technique is good only for certain hours. Also, because this method uses a different sample for each survey, there is bound to be some chance fluctuation from one sampling to the next. What appears to be a fairly dramatic change in a show's popularity may be only chance error.

Other researchers have used personal interviews in the attempt to discover what people have been watching recently. While this method does permit tabulation of viewing habits during odd hours, it ultimately depends on the accuracy of people's memories, and they can be woefully short. Well known shows can get overinflated ratings through this kind of study.

In order to counteract the problem, audience researchers have asked people to keep diaries of what they watch. Nielsen, in fact, uses this method to produce ratings for local television stations. The diary method, however, depends on people's willingness to keep track of their viewing. Though many people promise to do so, almost half fail to carry out the job. As a consequence, the information which is collected may not be really representative of the preferences of the population as a whole.

Moreover, the telephone, interviewing, and diary methods all share a fundamental weakness in that they depend on the respondents' honesty. Just as political pollsters often fail to locate hidden support for controversial candidates like George Wallace, some television viewers might not like to admit that they are really watching re-runs of "The Beverly Hillbillies." Then there are those people who seize the opportunity to manipulate the ratings. One young man whose family was sent a Nielsen diary says, "I just grabbed the thing and filled in all the programs I thought deserved a vote, whether or not we really watched them." Undoubtedly many people do the same.

The unique advantage of Nielsen's audimeters is that they do not depend on people's memories or honesty. The viewer just goes about his life as he normally would. The great virtue of the Nielsen system, however, is its greatest vice. The machine blindly records when the set is on, but it does not reveal who the actual audience is or how intently they are watching.

As pollster Lou Harris says, "The joke of it is that if somebody forgets and leaves the house with the set on, it's counted as if somebody is watching." The Nielsen is now rigged to flag any set which has been on for a continuous twenty-four hours, and a field representative is sent to see if there is an equipment malfunction. Reportedly this system once led a Nielsen worker to a home where the sample viewer had died watching television. Company spokesmen do not know whether the story is true, but if such an incident has not happened yet, it will inevitably occur sometime in the future.

The audimeter system thus can report only on set use, not on audience behavior. There are other serious drawbacks in Nielsen's methods which call into question whether his ratings should be given so much weight. The biggest problem may be that the company uses the same sample of twelve hundred again and again; people remain in the Nielsen sample for as long as five years.

The advantage of using a constant sample, a so-called panel, is that over a given period of time any change in the results indicates an actual change in viewer preference. A three-or four-point shift from one political poll to another may be only sampling error; by the laws of chance there will be some random variation from one sample to the next. Only if the people in the first sample were re-interviewed and said their views had changed could one be sure that the apparent change really reflected a trend.

Political pollsters seldom go back to the same sample, however, because they believe that the mere process of being interviewed can change people's outlook and opinions. Occasionally a pollster may return once to the same panel, but rarely more often than that, for the people will begin to think of themselves differently. Perhaps, embarrassed by their ignorance in the first interview, they will prepare for future ones. Lou Harris says: "I think you can go back to people a second time; you might get away with a third time. But after that, methodologically you're creating the risk of influencing your respondents. Especially if you're asking repeat questions, because they may well give studied responses." George Gallup has raised similar objections to using panels.

Nielsen has thus embraced a survey method which has been rejected, with good reason, by the political pollsters. He has done so mostly out of necessity; it would be very expensive to be constantly installing and removing audimeters in sample homes. In the final analysis, as Harris notes, "the question is whether the people in the panel are atypical because they know that they each represent sixty thousand households."

Harris says he has talked to Nielsen about this very problem. "What Nielsen is betting on is that people forget the devices are attached to the set and just go about their affairs as if nobody is watching. I suppose it may be different from going door to door and asking people questions, but I don't know if you could really prove it has no effect on people's behavior."

A number of journalists, who have wondered whether the Nielsen panelists are truly representative and how they regard their unusual responsibility, have tried to crack the web of secrecy which surrounds the entire operation. The Nielsen company is very nervous about possible tampering with its sample. With only twelve hundred households involved, it would not take much in the way of bribery to throw the ratings radically out of whack. As a result the firm refuses to let reporters talk to any of the Nielsen families. The families themselves have to sign an agreement in which they promise not to disclose their special status to anyone connected with the press or television.

Marvin Kitman, a columnist for *Newsday*, tried to circumvent these barriers by offering a cash reward to any Nielsen panelists who would consent to talk with him about their experiences. Kitman got a response, but it was from Nielsen's lawyers, who threatened the writer with a lawsuit for interfering with the company's contracts with its panelists. Efforts by *Life* and *TV Guide* were similarly thwarted.

Much to Nielsen's distress, *Los Angeles Times* reporter Dick Adler was able to succeed where others had failed. In a well-conceived and entertaining piece of investigatory journalism, Adler was able to both take much of the mystery out of the ratings system and demonstrate just how vulnerable to manipulation it is.

Adler described one Nielsen viewer as "Deep Eyes," in honor of "Deep Throat," the secret source who helped *Washington Post* reporters Robert Woodward and Carl Bernstein break the Watergate story. Deep Eyes was most conscious of his special influence, and he exercised it carefully. If there were a program which he thought particularly worthy, he would vote for it by tuning it in without actually watching it. "My one-twelfth of a rating point doesn't mean much in the national ratings, but we've had a lot of luck in the local Los Angeles market." There are only five hundred households in the Los Angeles market, so Deep Eyes' influence was particularly strong there. In his attempt to boost particular shows, he had his set on at least fifty hours a week.

Nielsen had chosen Deep Eyes to be in the sample not because of any personal characteristics nor through a random selection process, but because the house into which he moved was already wired for an audimeter. After reading Adler's account of how Deep Eyes deliberately used his set to manipulate the ratings, Nielsen officials undoubtedly cringed and regretted that they had not borne the expense of wiring a new house.

After Adler wrote about Deep Eyes, several other Nielsen panelists contacted the reporter. One of them was indignant that another panelist would abuse his franchise. Adler interviewed far too few people to determine whether most Nielsen people play the game straight. Even if they do, it does not take many like Deep Eyes to skew the results. Twelve households make a Nielsen rating point; eight constitute a share point. The life or death of a program can hang on a single rating point. Whether it gets that point may depend on the whim of people like Deep Eyes.

Nielsen spokesmen are quick to downplay the possibility that people in the sample may self-consciously try to manipulate the ratings, claiming that any such biases even out. Because the company will not let outsiders talk to all the panelists, it is impossible to know just how they do react to being watched. Nielsen says that the knowledge makes no difference, but that seems unlikely.

The Nielsen sample is probably inherently unrepresentative. Anyone who is in it is by definition atypical. There are other factors as well which make the sample biased. The company brags that the sample is a fair cross-section of the population and points to the fact that 28 percent of the panelists own Chevrolets, the exact percentage for the population as a whole. It does not mention, however, that blacks and other minorities have always been significantly under-represented in the sample.

Spokesmen for Nielsen say this disturbs them, but the condition has persisted from the outset. Lester Brown, in his book *Television: The Business Behind the Box*, suggests that apathy on the part of the networks and advertising agencies which sponsor the surveys explains why the situation has not been corrected. "So little valued has been the black man as a consumer of nationally advertised products that he was not properly represented in the Nielsen sample of the American television audience."

According to Brown, there was no outcry in the industry because the "ghetto Negro was not a target audience for most advertisers because, generally speaking, he was a low-income citizen with scant buying power." As a consequence it is likely that even though low-income people tend to be heavy television watchers, programs of special interest to blacks and shows aimed at urban viewers, like "Sesame Street," are probably under-rated by Nielsen.

There is still another source of sample bias, and this is one not only tolerated by the networks but welcomed by them. Nielsen's methods significantly overinflate the extent of television viewing in America. Not everyone who is asked to participate in the sample agrees to do so. Some people have understandable reluctance about being spied on, even if it is in a very limited way, while others simply do not want to be bothered. In such instances the company has to find a substitute household. Although the firm does the best it can to match the characteristics of the alternate with those of the household which declined, the very fact that one person says yes while another says no may mean that the one who agrees is more interested in television.

Bill Behanna, a Nielsen executive, says that their studies do in fact indicate that the alternates tend to be heavier users of television. "It is impossible to measure, but we will admit to the fact that ratings as a result are probably inflated to some extent."

An even greater source of inflation comes about from the way in which Nielsen tabulates the results for households with more than one TV set and for households for which the channel is switched in mid-program. Many homes have more than one set; if such a home is included in the Nielsen sample, each set is connected to the audimeter. If there are three sets in a given house and each is tuned to a different program, Nielsen credits each program with a household, instead of just a fraction of one.

This is even-handed in that each program is treated equally, but it does have the overall effect of inflating every program's ratings.

More and more American homes have two or three television sets, and often they are tuned to different programs. Ratings thus go up, but actual audience does not. At the same time that the number of televisions is expanding, the average number of viewers per set is correspondingly decreasing.

To compound this inflation, for a household to be credited to a given program it need not be tuned in for the entire duration. A mere six minutes is enough. Thus if a Nielsen viewer flicks on a set which happens to be tuned to NBC, goes back to the kitchen to make a sandwich, returns six minutes later and switches to CBS to see the program he really wants, both that program and the NBC program as well will be credited with a household viewing it. A person who watches only the first six minutes of a show (or any six minutes) counts as much in the Nielsen ratings as someone who watches it straight through.

Although the Nielsen system has been criticized for other shortcomings, few commentators have recognized the significance of Nielsen's strange penchant for counting fractions as wholes. The mystery of sampling and statistics apparently scares people away from challenging Nielsen's methods, but it really should not take a mathematician to see that if you are counting noses and count some noses twice, you end up with a bigger total.

Let us say there are three houses on a small street. In the first, the television sets are off; the Haynesworths are busy insulating the attic. In the next house, the Frontieros are enjoying an hour of reading aloud; their television is also quiet. In the third house, however, the Bells are watching TV. In stereotypical fashion, Dad is in the den watching the Pittsburgh Steelers make mincemeat of San Diego on ABC; Mom is looking at the NBC movie in the living room; and the kids have the portable upstairs so that they can see Archie Bunker call his son-in-law "meathead."

If the Nielsen Company were to survey the scene, it would credit each network with a separate household, even though television was actually being used in only one of the three homes. For this sample, each of the three programs would have a rating of 33 percent, which is supposed to mean that each one of the programs attracted viewers from a third of the homes with television.

Consider the same street a week later. Mom has already seen the movie, so she has gone over to help the Haynesworths, who have found that installing insulation is not as easy as advertised. The kids are over at the Frontieros' trying to get the characters straight amidst the many pages of *Nicholas Nickleby*. Only Dad is where he was before. After ten minutes of Howard Cosell, however, he switches to NBC, but after discovering that it is another movie with George Peppard, he turns to "All in the Family."

If Nielsen were to survey this scene, again each show would get a rating of 33 percent, because each program would be credited with a household viewing it. Hardly anybody on the street is watching television, but you would not know it from the ratings.

The inflationary effect of Nielsen's peculiar manner of counting is huge. In 1966 the Politz research organization telephoned twelve thousand households to check their viewing habits. Their survey showed 41 percent of the households using television in the evening. By contrast, Nielsen reported 55 percent. More significantly, the Politz survey found both fewer viewers per household and fewer adults watching prime-time television, 24 percent. Nielsen had adult viewership at 41 percent, well over one and a half times as much. *Life*, using an electronic device which detects the presence of a functioning television set, confirmed the Politz results.

Why, one might innocently ask, do the networks continue to sponsor the Nielsen ratings and attach to them so much importance, when they are probably so far out of line? The answer is that from the networks' point of view they are out of line in the right direction.

Network advertising time is sold on the basis of thousands of viewers delivered. Though there are all sorts of premiums and discounts, the basic price for a minute of prime-time advertising is two dollars and fifty cents per thousand viewers or, figuring roughly two viewers per household, five dollars for every thousand households.

To get a rough idea of the importance of a single rating point, calculate the cost of a sixty-second spot. If the program has a Nielsen rating of thirty, that ostensibly means that 30 percent of the country's seventy million households (the actual number is a little less) are tuned to that show—that is, twenty-one million households. The network could demand $105,000 for a minute of advertising time. Drop the rating a single point: the program loses 700,000 households and the network loses $3,500 of gravy on each of the six minutes an hour that it can sell. That's a loss of up to $21,000 an hour. Drop the ratings two or three points and the losses in revenue multiply accordingly.

Small wonder that the networks do not mind that Nielsen's methods of counting households inflate the ratings! It is one kind of inflation they can live with very happily.

Ratings are thus important because they are directly tied to money. A producer who can put together a highly rated show makes his network a pile of money. A producer whose shows bomb in the ratings gets fired. It is as simple as that.

The Nielsen ratings are reported down to decimal points; the decision to keep a program or the axe it can rest on a slim point or two. Even if the figures were totally free of bias, however, they really do not warrant this kind of respect. Nielsen's sample is about the same size as Gallup's and Harris' and consequently is subject to the chance sampling error of three points either way. Thus a program which has a rating of thirty may really be watched by as many as 33 percent of the households or as few as twenty-seven. Production costs are for the most part fixed, so that once a program earns back its investment, any additional revenue is virtually pure profit. A difference of six points, when translated into advertising revenue, can spell the difference between whether a network has a big winner or will lose money on a show.

Spokesmen for Nielsen concede that this range of error does apply to a one-time measurement of a show's popularity but contend that for a series, where the ratings are taken again and again, the sample becomes much larger and the chance error significantly smaller. This is fallacious. It would be true if each rating were based on a different sample, but Nielsen goes back to the same panel each time. If by chance that sample is a little biased in the direction of football fans, that bias will remain in the ratings and will not average out.

The use of a panel does make it easier to spot trends, that is, to determine whether a show's popularity is growing or flagging.

Contrary to what the Nielsen people claim, however, it in no way reduces the margin of sampling error. The rating for a given show is subject to plus or minus three points chance error (sometimes even more) whether that rating is based on one night or the entire season. When you read the ratings in the newspaper, however, there is never any hint that the numbers are subject to such a range of error.

If television has been less demanding in the quality of the ratings which firms like Nielsen produce, it has wanted more and more in the way of quantity. The audimeter system, for example, produces national ratings only. The sample of twelve hundred is far too small to be broken down into sub-samples to get ratings for individual stations; for instance, there may be only forty households in Massachusetts hooked up to the Nielsen computer. The company has set up special audimeter samples for the major markets—five hundred households each for New York and Los Angeles, and three hundred for Chicago—but it simply would not be profitable to set up similar systems in other areas. As a result, most television stations cannot look to the Nielsen audimeter reports to see how they are doing.

The audimeter system also fails to tell broadcasters and advertisers just who the audience is. The meter can record only that the set is tuned to a particular channel. It cannot reveal how old the viewer is and how much money he or she makes.

To determine local station ratings and to collect demographic data about television audiences, Nielsen conducts three mammoth "sweep weeks" each year. There may be as many as 190,000 households selected for each sweep week sample; the size is necessary so that there will be enough households in each local market, be it Boston or Birmingham, to constitute a workable sample. Each household is given a diary for every set it owns. The family is asked to record not just what programs are watched, but who watches them.

These sweeps are crucial for many local stations, as they often provide the only independent measurement of audience. If a station happens to do well during the sweep periods, it can charge higher rates for advertising for the other forty-nine weeks of the year.

The sweep weeks occur in November, February, and May, but you really do not need a calendar to know when they take place. Television programming during these periods clearly reveals that there is a crucial battle for audience going on. In May 1975, for example, NBC suspended its telecasts of the Monday-night baseball game and substituted higher-rated movies. Talk shows and variety programs vie for the best-known guests. Blockbuster movies and specials appear during the sweep weeks. When the Johnny Carson show operated out of New York, it would always move to Hollywood during the rating periods so that it could more easily get big-name celebrities. All of these are examples of a practice known as "loading" or "hypoing."

Federal Communications Commission regulations prohibit a station from engaging in "special promotional activities" during a rating period, but as of yet the government has taken no action against hypoing. For one thing, it is hard to prove conclusively that a particular program has been loaded. Who is to say that one comedian is a celebrity and another is not? Second, the networks, not the stations, are responsible for the programming, and the networks are not licensed by the FCC. It is common knowledge, of course, that the local affiliates push the networks into hypoing, as the networks themselves have nothing to gain by the practice.

Hypoing is another reason why television ratings are artificially high. According to Bill Behanna of Nielsen: "The complaint about hypoing really comes from those who are buying advertising time, not those who are selling it. If you use these reports to buy time, there is the contention that audience levels have been inflated and don't reflect normal programming. It's probably true." The end result is more advertising revenue for the affiliates.

There used to be another side of the coin. Until recently, Nielsen shut off its audimeters four weeks each year. During these so-called dark weeks networks could fulfill their public service duties by presenting political commentary, ballet, and the like without worrying about the ratings. Now, however, the audimeters are left on three hundred and sixty-five days a year, and the dark weeks are gone.

Nielsen's reliance on diaries for its sweep week information significantly inflates the ratings, perhaps as much as the networks' special programming efforts. Although the company puts diaries in a lot of households, it is lucky to get back 60 percent of them completed; sometimes the return rate is less. The people who take the trouble to log all their viewing probably care about television much more than the people who do not cooperate. Again the method produces erroneous data, but the error benefits the networks and stations which sponsor the survey.

The Nielsen Company is understandably self-conscious about its role as television scorekeeper. Often irate fans will write the firm and blame it for cancelling their favorite program. Nielsen sends such people a jaunty little booklet which says: "The answer —cross our hearts—is that we didn't. . . . All programming decisions are made by the network or the station."

Strictly speaking, of course, Nielsen is right: it merely counts the house; it does not close the show. But since the count is what determines whether or not the show goes on, the work Nielsen does is really decisive.

Like the political pollsters, the rating people invoke the democratic ideal to justify what they are doing. Press spokesman Bill Behanna says, "The broadcasters are looking for a mass audience, and the name of the game is to get all of the people all of the time." This outlook apparently does not bother Behanna. "The rating system is probably far more democratic than the United States Congress." In essence, Nielsen says that people get what they vote for.

The notion that ratings are democratic assumes that people have a wide range of choice, but this is not more true in the case of the three networks than it is with the "big three" automobile manufacturers. You get to choose from what the networks let you see.

New York Times television writer Lester Brown has stated: "One of the myths about American television is that it operates as a cultural democracy, wholly responsible to the will of the viewing majority in terms of the programs that survive or fade. More aptly, in the area of entertainment mainly, it is a cultural oligarchy, ruled by a consensus of the advertising community."

According to Brown, the key to programming is density of viewership, particularly among the middle class, which buys most of the air freshener, aspirin, and automobiles. Programming clusters around the middle, with the networks playing Tweedledum and Tweedledee. Brown says: "This emphasis on the popularity of shows has made television appear to be democratic in its principles of program selection. In truth, programs of great popularity go off the air, without regard for the viewers' bereavement, if the kinds of people it reaches are not attractive to advertisers."

Judging from the fare which television usually produces, it is clear that the ratings competition does not breed better or more diverse programs. The networks crawl all over each other fighting for the center of the population, when everyone—viewers and broadcasters alike—might be better off if their networks had distinct identities and pitched their programs to particular segments of the community. Only then would there be real choice.

Thanks to the ratings, however, we now get happy news on all three channels. One anchorman wears a three-piece suit, another has blue eyes, and a third parts his hair in the middle. "There, see," says television proudly, "it is a democracy. Take your pick!"

11

Prophets and Losses
Commercial Polling

Political polling is only the tip of the public opinion research iceberg. The great bulk of polling is done for private businesses. United States companies may spend as much as half a billion dollars a year, perhaps more, for outside market research and attitudinal studies. Though George Gallup and Louis Harris are best known for their syndicated newspaper columns, that work makes up only a small fraction of their activities. The great majority of their revenue comes from private corporate clients.

This commercial surveying is less conspicuous than political polling, but it too has disturbing implications. It affects the products we buy, how much we pay for them, and how they are advertised. Just like the political polls, commercial surveys can be rigged and manipulated. Rather than making business more responsive to the needs of consumers, they often make it less so.

Political and commercial polling often go hand in hand. Most political pollsters do commercial work; those who publish polls, whether nationally or locally, exploit the publicity they get from them to attract corporate business. In fact, every survey which Gallup and Harris conduct includes scores of questions tacked on for corporate clients. Thus a person who is interviewed may have to rate Henry Kissinger in one question and evaluate different brands of peanut butter in the next.

Gallup's base charge is about three hundred dollars for a single question presented to one segment of the sample. A series of twenty-five questions asked of all fifteen hundred people interviewed might cost as much as twenty thousand dollars. This "piggy-backing" is very lucrative for the pollsters, because the per-question charges are much higher than when a complete study is underwritten by a single client.

A number of pollsters question whether businessmen who shell out for piggy-backing really get their money's worth. Gallup makes it a practice to ask his newspaper questions first, while people are fresh and eager to participate. It follows that they must be somewhat stale when they come to the tacked-on questions. There is also the criticism that a few questions simply cannot give a useful reading of public opinion. Political pollster Pat Caddell suggests that the practice of tacking on questions can also corrupt the rest of the process. "It can become a case of the tail wagging the dog."

Even where a survey is totally devoted to a single corporate client, many of the techniques are identical to those of newspaper and political polling. Some specialized business work, of course, does call for innovations.

The most basic kind of commercial polling, and the kind which has the oldest roots, is market research. Many of the pioneer public opinion pollsters got their start by testing the markets for various products; from there they branched into other fields. The career of the late Elmo Roper was typical of this pattern. Roper began not as a researcher but as a salesman for a jewelry manufacturer; he sold its line to retail stores.

One of Roper's functions was to pass judgment on possible items for the coming year's line. When he was first hired, he was very impressed by his colleagues who could simply look at something and say whether or not it would sell. He did not have the confidence or the experience to make decisions of that sort, however, so he made it a practice to take samples out to jewelers he had sold to in the past and ask them what would move and what would not. He would then return to his home office and tell the company which items should be pushed and which dropped.

This was market research in its crudest yet perhaps most useful form. Roper quickly had the best track record of all the salesmen. As his son, Burns Roper, puts it, "He wasn't shooting from the hip, just saying what he himself liked or disliked." The senior Roper's success convinced him that the routine he followed could have a much broader application. He began doing similar surveys for other businesses with different kinds of products. At the outset, he did most of the interviewing and worked on an intuitive, subjective basis.

Roper's clients began to ask that people's preferences be broken down according to their age, sex, or geographic location. This required more work, so Roper had to hire other people to do the interviewing, which in turn mandated written questionnaires. Methodology grew out of need and was progressively refined. At the core, however, Roper's function was to determine who would buy what.

There is nothing pernicious about this kind of market research; in fact, it is very useful in any complex economy. Whether one has great faith in the efficiency of the free enterprise system or favors a planned economy, there is a need for information which will help supply meet demand. Good market research can tell a manufacturer that a certain product just is not wanted and thus save him, and society, the cost of its production. Likewise, if demand is higher than anticipated, good market research can tell a company to produce more and thus take advantage of economies of scale.

Business now turns to pollsters to find out not only what it will sell, but what it has already sold. The principal activity of the A. C. Nielsen Company is not rating television (that constitutes just a seventh of its revenue), but a service called the Retail Index which monitors the movement of food and drug products at point of sale.

More than two-thirds of Nielsen's income derives from knowing, among other things, how many bottles of ketchup were sold in the last two months. Surprisingly it is the people who make the ketchup who pay to get that information! According to a Nielsen spokesman, "It may sound funny, but the manufacturer doesn't always know what its own sales are. It knows over a long period of time, but we can provide a much faster reading of what's happening in the marketplace."

Back in the 1930's, when Elmo Roper was developing his market research techniques, other pollsters like George Gallup and Archibald Crossley were exploring the same field. The success they all had in predicting Franklin Roosevelt's re-election in 1936 put them in the public spotlight, and their commercial polling took off.

Gallup discovered that he could use market research not only for products, but also for ideas. One of his organizations, Audience Research, tested story ideas for movie producers. Producers like David Selznick, Sam Goldwyn, and Walt Disney, among others, paid Gallup to try out the plots of hundreds of movie scripts on a representative sample of the public. Gallup says, "We found a way to test story ideas, and when other important factors were also measured and taken into account, such as star values, titles, publicity penetration, and the like, we could make surprisingly accurate estimates of the box office gross for any given picture."

For the most part, straight market research has provided reasonably accurate data for business, particularly when it has been gathered by experienced pollsters. There have been some instances, however, where the market researchers have been spectacularly wrong. In the mid-1950's, for example, Ford Motor Company used surveys in formulating plans for a new line of cars to compete with Oldsmobiles and Buicks.

In April 1957 market researcher Albert Sindlinger, who specializes in automative studies, reported to Ford that the public would buy two hundred thousand units the first year the car came out. According to Sindlinger, the Ford executives were jubilant when they heard this news. The company went full steam ahead with plans to produce two hundred thousand of the new cars. They were to be named Edsels.

The car was introduced in September of that year. By then, however, the economy had taken a dip, and Sindlinger revised his estimate of the market down to fifty-four thousand. Sindlinger says that the Ford executives were not at all happy with his gloomy prediction. "I was physically thrown out of the office." Ford was deeply committed by that point and felt that it could not turn back. Sindlinger's estimate, however, turned out to be almost exactly right. Ford ultimately scuttled the Edsel. Almost twenty years have passed since then, but the company still has not lived down its mistake.

In spite of such incidents, market research has become a fundamental element in business planning. Few companies, large or small, introduce new products without first testing the waters with elaborate surveys. Ironically, market research has become such an accepted tool that many companies now support their own in-house research staffs. Advertising agencies and management consultants have also gotten into the act. As a result, in the past ten years some pollsters who relied solely on traditional market research have lost so much business they have had to fold. Only those pollsters who have been able to sell more sophisticated services have really prospered.

Specialization has taken a variety of forms. Some firms, for instance, specialize in researching the effectiveness of advertising. To cite an example, several years ago Alan R. Nelson Research of New York did an exhaustive study of the value of various athletes' endorsements of products. The poll, called Sports Insight, tested almost two hundred athletes on a variety of scales, such as how well they were known, admired, and liked. It revealed that notoriety and popularity can be two quite different things. At the time, Muhammad Ali was the third best-known athlete, but ranked last in personal liking.

The poll was underwritten by twenty national advertisers who each paid six thousand dollars to get the results. They included Manufacturers Hanover Trust, ABC, and advertising agencies for Volkswagen, Budweiser, and Coca-Cola. If a sponsor is going to shell out thousands of dollars to get an athlete to plug his product, he wants to be sure the endorsement will help, not hurt.

The results of the poll were widely distributed, but apparently not everyone believed them. Joe Namath, who ranked second in familiarity, supposedly was only one hundred and fifty-sixth in terms of the overall value of his endorsement. Namath can laugh all the way to the bank, however, as none of the hundred and fifty-five athletes who finished above him make the kind of money he does from endorsements. In 1975 he signed a contract with Fabergé which would reportedly earn him five million dollars. On the other hand, Schick paid Mark Spitz an enormous amount to endorse their products exclusively, yet they have made little use of him. Perhaps polls like Sports Insight, which ranked his endorsement near the bottom, convinced them that they were better off swallowing their loss than being too closely associated with a negative personality.

Surveys are also used to test the names of new products or companies. Several years ago, for example, Daniel Yankelovich's firm was purchased by Leasco, a large conglomerate. Soon afterward, in part because of a Yankelovich study, the parent company changed its name to the more impressive sounding Reliance Group, Inc.

While some pollsters specialize in advertising research, others have found that they can sell surveys of special groups. Louis Harris, for example, has an Executive Interviewing Force which can be sent out to gather the opinions not of the proverbial man in the street, but of business leaders, congressmen, and high-level bureaucrats. Harris does not usually trust his ordinary Harris Survey interviewers to do this kind of work. The executive interviewers are often men, often well-educated, and sometimes former executives themselves.

The Roper Organization also offers surveys of the elite. Its president, Burns Roper, recalls: "A client once asked us to interview a sampling of Nobel Prize winners, senators, and former United States presidents. I think it worked out to five hundred dollars per interview."

Still other pollsters have found it profitable to concentrate their surveys in particular areas. Albert Sindlinger, for example, has specialized in economic forecasts, polling people continually on their present financial situation and their expectations for the coming months. By asking people what they earn, whether they expect to get a raise, how they view conditions in their locality and the like, Sindlinger comes up with an index of consumer confidence and money supply. Relating that index to supplementary information, Sindlinger says it is possible to forecast such things as new car purchases, bank loans, and stock market activity. He has a number of subscribers who apparently agree, as they pay up to five thousand dollars annually for the privilege of subscribing to his weekly service.

Private pollsters are discovering that they can sell their services not just to businessmen, but to lawyers as well. Some attorneys have used polls to make strategic decisions. Boston lawyer William Homans used a poll in defending Dr. Kenneth Edelin, accused of manslaughter in performing an abortion. The survey convinced Homans that a change in venue would not substantially improve his chances of getting a sympathetic jury. Certain things turned up by the survey—such as the fact that 14 percent said they would vote to convict no matter how the judge instructed them—persuaded the court to allow much more extensive examination of prospective witnesses than is usually the practice.

Polls have also been used as evidence. A New York lawyer, defending a man accused of violating the obscenity laws, introduced a poll on local attitudes toward sexually explicit material. One of the legal tests of obscenity is whether the material in question is patently offensive in terms of contemporary community standards; the lawyer commissioned a sixty-five-hundred-dollar survey to show that his client's publication was within accepted bounds.

The most conspicuous and controversial use of polls in the courtroom has been in jury selection in political trials. In the 1971 trial of the Harrisburg Seven, two sociologists, Jay Schulman and Richard Christie, conducted a sweeping survey in which they sought to correlate demographic factors—such as age, sex, and occupation—with sympathy toward the defendants. Lawyers have always employed rules of thumb in choosing jurors, but this was the first real attempt to transform that crude art into a science. The defense team used the poll to guide their questioning of prospective jurors, and judging from the result—a hung jury, with ten out of twelve voting to acquit —they chose wisely.

Schulman and Christie have become jury-selection specialists, using their survey techniques in several of the Attica trials, the trial of the Wounded Knee leaders, and the recent trial of Joann Little. Thus far they have chalked up an impressive record. These techniques are by no means the exclusive domain of the left. In the New York conspiracy trial of former Nixon cabinet members John Mitchell and Maurice Stans, the defense similarly took a poll to discover the characteristics of an ideal juror.

Not everyone in the legal profession welcomes this growing use of polls. In criminal cases, they add to the already high cost of defense; only defendants in celebrated cases can now afford them. There is even greater cause for concern when polls become the test of substantive issues, such as obscenity. Our First Amendment rights would be shallow indeed if some book or film could be censored simply because a poll reported that it was offensive to a majority of citizens.

The most prestigious, and the most lucrative, polling speciality has developed in just the last few years. It is the syndication of unusually intensive studies of public attitudes toward issues especially important to business. The subscribers are not newspapers but corporate executives, and the membership dues are high. The pooled revenue allows a pollster to conduct a much more sophisticated survey than he could do for any individual client.

The topics for study are almost always chosen by the pollster, not the subscriber. One survey may explore public reaction to energy problems; the next in the series might deal with opinions about tax policy; a third could consider public attitudes about work. All the studies are intended to give the executive a better sense of how consumer and political pressures are likely to affect his business.

Only the best-known pollsters have been able to sell such services successfully. In essence, the pollster has to build his reputation before he can command the kind of price these services sell for. The Roper Organization issues its *Roper Reports* ten times a year to forty-five subscribers, each of whom pays thirty-six hundred dollars for the basic service. This one enterprise thus grosses Roper more than $150,000 annually.

According to Burns Roper, the subscribers are not exclusively industrial corporations. "We've also got some government agencies; we've got the Democratic National Committee, all three networks, some advertising agencies, and a couple of newspapers." Subscribers can release the data on a selective basis, such as in annual reports or speeches by company executives.

The Roper service is considerably cheaper than its competition. Louis Harris syndicates his *Harris Perspective* to forty firms at twenty-five thousand dollars a year, which grosses an even million dollars for his own company; other pollsters estimate that 25 to 30 percent of that is profit for the Harris organization. According to Harris, the *Perspective* produces more than ten times the revenue than does the much better known Harris Survey.

Harris has implied to those in the business press that subscribers may sign up for his service only at his invitation. This apparently is intended to give the enterprise an air of exclusivity, like a private club, but the fact is that if you have twenty-five thousand dollars for the dues, your application for membership will be automatically accepted.

Opinion Research Corporation of Princeton, New Jersey, offers the *Public Opinion Index*, a battery of twenty-four surveys a year. Fifty subscribers, including five of the six largest industrial corporations in the country, pay an annual fee of six thousand dollars to receive the *Index*.

Political pollster Pat Caddell professes disdain for market research, but he has not been able to turn his back on the mother lode. His firm, Cambridge Survey Research, produces the *Cambridge Reports*, but, perhaps because of its close ties to liberal Democrats, its service thus far has not fared as well in the business community as have most of the others.

Daniel Yankelovich has been the most successful pollster in this sort of venture, even though in some respects he seems an unlikely candidate for that distinction. His firm offers two sophisticated services. One of them, the *Yankelovich Monitor*, is a once-a-year report on changing American values and life styles; eighty subscribers pay $11,500 each to read that single survey! At twice that price, twenty-five subscribers get *Corporate Priorities*, a study of public attitudes toward the responsibilities of business in respect to the quality and safety of products, industrial pollution, and the like. Yankelovich puts on special seminars for executives of the subscribing corporations, which include General Motors, Du Pont, and A. T. & T. Yankelovich's firm grosses almost a million and a half dollars a year on the two surveys.

Yankelovich's success is certainly warranted in light of his experience. He founded his firm in 1958, and for many years it has been a leader in market research. His career is the reverse of the usual pattern, in that he became fully established in the commercial world before he ventured into newspaper work; he was a success long before he began polling for the *New York Times* and *Time*. Though still not as famous as Gallup or Harris, Yankelovich has been well respected within the polling profession for a long time.

What makes Yankelovich's eminence a bit surprising is his

demeanor, which is anomalous. He is a fairly short, middle-aged man; he is balding, and, perhaps as compensation, he has a rather full mustache. Instead of looking like a captain of the polling industry, he looks as if he would be far better cast as a sociology professor in an urban university.

His background is considerably more academic than those of most of his rivals. He went to Harvard College, then stayed on to do graduate work in social relations; he completed his dissertation at the Sorbonne, but while most pollsters flaunt their credentials, Yankelovich is self-effacing. He does have the title "Doctor," but he does not like people to use it. "It makes me uncomfortable. I think I should pull out a stethoscope."

Yankelovich's work is generally conceded to be the most sophisticated in the business, and this no doubt reflects his scholarly training. He is an enthusiastic proponent of survey research, but, more than most other pollsters, he also appreciates what polls can and cannot do. Where Yankelovich stands apart is in his ability to formulate an imaginative series of questions and interpret the answers.

There is no question that Yankelovich's syndicated surveys are of the highest quality, but why is it that big business is willing to pay so much to get them? How is all this sophisticated information used? No doubt part of the appeal lies in the mystique which the big-name pollsters have been able to create. No matter how important an executive may be within his own company, he is bound to feel a little more so if he can call up Lou Harris or Dan Yankelovich and get an up-to-date reading on the public pulse.

The environmental movement, consumerism, disenchantment with gas-gobbling cars all caught businessmen off-guard. The pollsters are promising that, for a price, they will not let this happen again. All the poll-takers who syndicate special services echo this theme. Harris refers to his service as an "early-warning device." Roper says: "Businessmen have been surprised too many times. I think they want to have their antennae out just like senators and congressmen so they can anticipate instead of reacting to things when they are suddenly on you." Yankelovich concurs: "Our services involve the changing social responsibility of a corporation, the changing expectations and rules of the game."

If these surveys really encouraged corporations to conform to changed expectations, then we could all be very thankful. Unfortunately the business world, confronted with news of social unrest, has responded by summoning all its creative and financial resources—not to re-order priorities or correct old deficiencies but to produce an avalanche of expressions of corporate concern. Four-color magazine ads show herons nesting in oil wells. On television, trim executives wearing hard-hats declare, "At U.S. Steel, we're concerned." Mobil tells us that it warned of an energy crisis long ago, but no one listened. Atlantic Richfield wants our suggestions on mass transit.

The polls have encouraged business to see public restlessness as merely an image problem. Television and magazine ads now stress fuel economy and control of pollution, but what actually has changed? Gasoline prices still mysteriously rise in unison. Automobile manufacturers seek to postpone anti-emission regulations. The pollsters report on disturbing public attitudes, and business responds by trying to change the attitudes, rather than the practices which caused them.

Whether the boom in syndicated polling services will continue remains to be seen. To the extent that it depends on corporate paranoia about anti-business sentiments, the pollsters can create their own market. So long as the news they present is bad, business will feel compelled to buy it. When executives see Opinion Research surveys which show that out of thirteen institutions in the United States, "large companies" are least respected by the public, the business leaders have to know more. The pollsters are working both sides of the street: they make a grim diagnosis, then they sell the cure.

The quality of the best syndicated work, such as that by Yankelovich, is least equivalent to that done in academic circles. It may be the best public opinion research conducted. The quality of the more conventional commercial work is spotty. As in political polling, when the pollsters are collecting simple facts they are on firmer ground than when they seek out opinions. For example, any pollster ought to be able to come close to estimating how many cars people own and how long they keep them. This is not to say that the pollsters are infallible in collecting factual data. Some people exaggerate their incomes, while others may forget just what brands they do buy. One company was sufficiently skeptical about people's description of their purchasing habits to send its interviewers not door to door but garbage can to garbage can, in order to see what people really were consuming. They found that people drink two and a half times the liquor that they tell interviewers they do. Research firms have also used hidden cameras on buses and subways to watch how riders respond to transit ads. Even without rummaging through refuse or spying on people, researchers can usually come up with fairly accurate factual information.

When the surveyors get into attitudes, however, they run into the same obstacles that bedevil polls on political issues. Opinions are inherently elusive and do not fit neatly into "yes," "no," and "undecided" categories. It can be difficult to measure the intensity of attitudes. When one person says he feels "very strongly" about industrial pollution, he may mean something different from someone else who uses the same words.

There are some pollsters who believe that these problems are handled much better in some commercial polls than they are in the typical newspaper survey. Peter Hart, who principally does private political polls, says: "The stuff that Harris and Yankelovich do for the newspapers isn't nearly as good as their private work. Yankelovich has done some brilliant studies with questions I never thought of; the same is true for Lou. But you never see that quality in the newspapers."

To the extent that Hart is right, quality is a consequence of two factors. One is the money available. Both Yankelovich and Harris have much bigger budgets for their commercial studies than for their published polls; this allows them to pre-test questions, interview more intensively, and spend more time analyzing the results.

Moreover, even if a newspaper pollster had an unlimited budget, he still would be up against severe space constraints in presenting his results. Harris' column runs seven or eight hundred words and is usually edited down to a fraction of that. By contrast, there are no comparable limits for private surveys; some reports are the size of coffee table books. Issues can be explored at great length. Indeed, Irwin Harrison, a Boston pollster, thinks that many private polls today are too long. "I suppose it looks impressive to have page after page of print-out and analysis, but it can get so unwieldy that the client doesn't know what it all means."

Though most pollsters feel that commercial surveying is of high quality, Pat Caddell dissents. "Imagination is not the hallmark of most market research that is done in this country. It's primitive compared to the kind of work that's done politically. Light years behind." Caddell says that he has no interest in doing straight market research and emphasizes that his subscription service, *Cambridge Reports,* deals with "the really interesting public problems of corporations, which require analysis of attitude data."

The biggest threat to the accuracy of commercial research is not technical problems but the client himself. Sindlinger says that he learned this the hard way, when he did the polls for Ford Motor Company on the ill-fated Edsel. "When I told them in the spring that they were going to sell two hundred thousand Edsels in the first year, they couldn't have been happier with my work. They didn't ask me a thing about my methods. But when I came back in September and cut the forecast down to fifty-four thousand, then they climbed all over me, challenging every decimal point."

This experience and others convinced Sindlinger that "people hire private pollsters to prove a point—no one's going to pay someone to prove they're not number one." According to him, many businessmen determine the course of corporate policy and then commission a poll to justify their decision. If their action does not pay off, they can always make the excuse that it was based not on their own judgment but on solid research.

Sindlinger says, "If you're halfway skillful in opinion research, you can in the same interview prove both sides of any subject by the way you structure the questions." With hundreds of polling firms, large and small, throughout the country, a client should have no trouble finding one which will produce data to his liking.

Sindlinger cites two examples of corporate closed mindedness about polls; both involve the automative industry. "All through the sixties, my surveys were showing a tremendous market for truly small cars. The Big Three just refused to believe it. They said the only people who would buy Volkswagens were college professors in Cambridge and Berkeley." The manufacturers' own in-house research staffs fudged their surveys, according to Sindlinger, to conform to the company line. If they had reported the truth, he says, they would have been fired.

More recently, in the early seventies, Sindlinger warned that more and more people were going to stick with their present cars. New car sales were bound to drop. General Motors refused to believe it and commissioned a major advertising agency to conduct a study to find out why people buy cars. The agency reported back that sales resistance could be overcome. How? Through massive advertising—that's what their poll said sold cars. This was just what G.M. wanted to hear, and the advertising agency, of course, was delighted to say it. The report proved wrong. In spite of massive discounts, cars remained unsold, and hundreds of thousands of auto workers had to be laid off.

George Gallup does not agree with Sindlinger about much of anything, but he too sees a strange willingness on the part of people in business to deceive themselves. Gallup says that on occasion advertising agencies have been able to learn the sampling points for some of the Nielsen television surveys. The agencies would heavily promote their clients' programs in those areas in order to boost the ratings. Higher ratings mean stiffer time charges by the stations and networks, but, according to Gallup, "The agency and, strangely enough, even the advertiser wanted to be kidded about the size of his audience." Gallup's example is another bit of evidence for Sindlinger's proposition that from the client's point of view, the most important thing about any survey is that it must show him out in front.

So long as the pollster gives the client what he wants, there is rarely any dispute about the quality and reliability of commercial polls. In 1975, however, there was a heated battle about the validity of magazine audience surveys. The connection between pollsters and journalism is deep. Gallup, for example, began his work by devising methods for testing newspaper readership. Leo Bogart, a past president of the American Association for Public Opinion Research, has done extensive studies of the popularity of various newspaper features.

Several research firms specialize in magazine audience studies;

in many respects they are to the print media as the A. C. Nielsen Company is to television. Advertisers want to know just how many people see various magazines in order to figure where they can most effectively buy space. Circulation alone is not a guide to readership, as many magazines are passed along to other people. This pass-along readership does not help the publisher in terms of subscription or newsstand receipts, but it is a critical factor in advertising revenue.

Advertising rates are principally based on paid circulation. If two magazines have identical circulation, and thus comparable rates, then advertisers will look to measurements of total readership to see which one will give them the most exposure for their advertising dollar. Currently a billion and a half dollars are spent each year on magazine advertising.

Magazines once commissioned their own audience studies, but it became increasingly obvious that these were self-serving and wholly unreliable. As a result a number of syndicated surveys grew up; because these syndicators were sponsored by many magazines, they seemed to be free from bias. Just as the need for one authoritative scorekeeper made Nielsen dominant in television ratings, so did one firm, W. R. Simmons & Associates Research, come to be the most important force in magazine research. Indeed, an attempt by Nielsen to move into the field got nowhere.

During the 1960's the advertising industry considered the Simmons figures to be gospel, yet Simmons' very success may have been its undoing. The company took on more and more subscribers, and in order to elicit comments about scores of magazines and countless products its questionnaires had to run almost a hundred pages. Some researchers began to doubt that the people who were interviewed could remain interested through such a massive interrogation.

Competition from a new firm, plus growing industry concern about methods, prompted Simmons to change some of its practices in the early 1970's. As an apparent consequence, the firm got embroiled in a controversy which has greatly damaged the credibility of magazine surveys.

For years *Time* magazine has had a paid circulation about 50 percent larger than that of its rival, *Newsweek*. Simmons consistently reported that *Time*'s total readership was also roughly 50 percent larger. In 1974, however, Simmons got into deep trouble when it reported that while *Time*'s audience had dropped about 13 percent, *Newsweek*'s had dramatically increased to a point where it almost equalled that of *Time*.

The report was devastating for *Time*, because it ostensibly meant that its advertisers could reach an audience of comparable size at far less cost by switching to *Newsweek*. When the figures were announced, it was estimated that *Time* could lose four million dollars or more in 1975 because of the lower ratings.

Time responded by suing Simmons. All sixty of the magazines whose audience is measured by Simmons pay for that privilege, just as the networks pay Nielsen. *Time* sought a declaration by the court that it did not have to pay the $188,000 bill outstanding to Simmons on the ground that the controversial study had been based on "biased and unreliable statistics." *Time* said that if the most recent study proved to be accurate, then it was suing for payments for earlier studies which had shown that *Time*'s total audience was proportionately larger than *Newsweek*'s. In essence, *Time* was arguing that either the most recent study or the ones which preceded it had to be wrong.

In bringing the suit, the magazine alienated many of the advertisers and agencies it was desperately try to keep. Some people in the field feared that *Time*'s muscle could silence independent auditing of magazine readership. Indeed, several months after the suit was filed Simmons announced that it was suspending its research, supposedly to concentrate its resources on fighting the suit. Frank Stanton, the chairman of Simmons, stated, "It is obvious that research auditing firms cannot operate under circumstances where they must provide acceptable findings to powerful and well-resourced interests in order to stay in business."

The real reason for suspending its survey may have been tactical, for the *Time* suit put Simmons in a bind. If the next survey again showed *Time* and *Newsweek* with equivalent audiences, that would cast greater doubt on the earlier results. But if the new study showed *Time*'s audience to be larger than *Newsweek*'s, that would be ammunition for *Time*'s claim that the controversial survey was indeed erroneous.

Within months of the *Time* suit, *Esquire* went to court against

Simmons, not because Simmons reported that *Esquire* had lost 44 percent of its audience, but because Simmons had not included it in its forthcoming survey. This omission, *Esquire* claimed, implied that it was no longer significant in the magazine market. During the same period, *Girl Talk* sued Axiom Market Research Bureau, a competitor of Simmons, for more than a million dollars, which the magazine claimed was the amount of advertising revenue it had lost as the result of an Axiom survey which indicated that its audience had dropped by more than a half.

Time and *Esquire* ultimately settled their respective suits with Simmons, and *Girl Talk* withdrew its action against Axiom. The magazines now contend that they won assurances that better methodology will be used in the future, but many in the advertising business believe that they really got very little in the way of concrete concessions. In any event, by calling the Simmons figures into question, *Time*'s suit did blunt their impact, and the magazine lost less ad revenue than had been anticipated. That the glaring inconsistencies in survey data were brought to light has, at least for a time, made the magazine and advertising industries much more skeptical about these studies.

Although the lawsuits never went to trial, they are bound to set a precedent of sorts. Before long, we may see a television producer who is unhappy with his ratings sue A. C. Nielsen, or a potential presidential candidate bring Gallup or Harris into court.

There are times, of course, when reliable data is the last thing that a client wants. Businessmen, like politicians, will pay for phony polls which they can wave around. In 1975 the Business Roundtable, a group of powerful business executives, underwrote a twenty-five-thousand-dollar survey by Opinion Research Corporation. The poll was released during the height of congressional debate on the creation of a federal consumer protection agency. According to the survey, 75 percent of Americans were opposed to the creation of such an agency. Senator Charles Percy, a sponsor of the proposal, submitted the survey to a polling expert at the Library of Congress who concluded that the poll was blatantly loaded.

The key question was, "Do you favor setting up an additional

Consumer Protection Agency over all the others, or do you favor doing what is necessary to make the agencies we now have more effective in protecting the consumer's interests?" By characterizing the agency as "additional" and implying that other agencies were capable of meeting the problems of consumers, the pollster had steered the issue away from consumerism and toward bureaucracy. Just to be sure the respondents were thinking about the perils of big government, the pollster preceded the question with fifteen others about federal agencies. The overall effort was to create an anti-government attitude, and, judging by the final percentages, it was successful.

Opinion Research Corporation does a lot of polling for Republican candidates. The firm is well known in the polling profession for its willingness to let its political clients use its name to legitimate selective and misleading leaks of its surveys. Apparently it accords the same privileges to its commercial clients. If a public interest group had come to O.R.C. with twenty-five thousand dollars first, it probably would have produced a poll showing Americans not opposed but in favor of creating a federal consumer protection agency by a three-to-one margin. The same is probably true of many other pollsters.

The businessmen who bankrolled the O.R.C. poll must have figured that it was a good investment, knowing how uncritically most of us regard public opinion surveys. If there is a number attached to a piece of information it must be right.

There are some surveys which, though crudely done, have acquired the status of an institution. Every week readers of the *New York Times* Sunday book review section flip to the back to see who is moving up and down on the best-seller list. Inclusion among the select ten is an author's certification of success. To get the number one spot is unquestionably to be at the top of the publishing world. Most people mistakenly assume that the list is a precise tabulation of all the books sold the week before, but, oddly enough, that information simply is not available.

The *Times* cannot rely on publishers to report accurately on what's selling. Publishers (and writers) are notoriously careless with the truth when it comes to talking about how many books they have sold. It is very tempting to inflate sales figures in the hope that they will become a self-fulfilling prophecy. Even if the

Times could carefully monitor shipments to bookstores, shipment figures would not necessarily be revealing, because the stores have the right to return the books if they are unsold.

By default, the *Times* must turn to the bookstores themselves, yet they seldom keep an up-to-date count on what is moving. The *Times* therefore does not get actual sales totals, merely the store managers' impressions of what titles are selling; they are ranked in no particular order. The best-seller list is based on people's memories, which are undoubtedly faulty and impressionable: reports from managers may well be affected by heavy promotion or even the previous week's best-seller list. A *Times* spokesman concedes, "At best, it's only a guesstimate."

The *Times* surveys only general bookstores, so a title in a specialized area can do extremely well yet never be reported. Billy Graham's latest book, *Angels,* apparently sold hundreds of thousands of copies in religious bookstores but was not recorded on the *Times* list until it started moving in general outlets as well. In recent years the *Times* has expanded its sample, but some publishers feel that it still does not accurately represent national sales. A biography of Bear Bryant, the Alabama football coach, sold very well in the Southeast, but the book never made the best-seller list, even though some books which actually sold fewer copies did.

The *Times* insists that its survey cannot be manipulated, but if someone is sufficiently determined, it may be possible to manufacture a best seller. Chuck Barris, the author of a recent novel, *You and Me Babe,* was so dissatisfied about the way his publisher was failing to promote his book that he decided to sink tens of thousands of dollars in buying it up in those stores known to be in the *Times'* sample. The purchases put the book on the best-seller list, which then generated enough orders from other stores to get it selling in its own right.

Once a book makes the best-seller list, it is likely to stay there for some time. During 1975 fewer than forty works of fiction made the *Times* list; there was not much more movement among non-fiction titles. Given the great impact that the list has on sales, the *Times* should be much more candid about how crude a tabulation it really is.

Commercial polling thus includes a whole range of activity, some of it legitimate and useful, some of it not. The past ten years particularly have brought major changes in the field. Pollsters have lost much of the straight market research work to in-house staffs. Audience survey firms have been threatened by lawsuits. But for those pollsters who had the foresight to sell specialized subscription services, the past decade has been lucrative.

It remains to be seen whether the subscription services of Roper, Harris, Yankelovich, and the rest will continue to prosper, or whether in time their novelty will wear off. So long as big business is nervous about public opinion, executives will feel a need to be in the know, but whether they will be willing to pay tens of thousands of dollars for information which may be available through other channels is another matter. In any case, however, if the past is any guide the pollsters will fall on their feet and find new ways to peddle their services.

12

🦉🦉

Policing the Polls

Public opinion polls determine who runs for political office and, often, who wins. They also influence the tone and content of debate on great national issues. Television, the mammoth industry which is the source of most of our news and is also our national pastime, is wholly at the mercy of the polls. Commercial surveys are important in every phase of the marketing and promotion of the products we buy.

All of these polls are subject to error and manipulation. Even polls which are published in good faith can be dangerously misleading, and of course not all polls are legitimate.

In spite of their great importance, public opinion polls hover strangely apart from other institutions. Not contemplated by the Constitution, nor regulated by Congress or the courts, nor checked by the press, the polls have multiplied in number and influence. Nothing presently exists to keep them in balance.

Given all the pitfalls in polling, given the potential harm that polls can do to our political process, why is it that the pollsters have not been corralled? Is there any way that the polls can be regulated and the power of the polls be brought under control?

Unfortunately, most of the cries for reform have come from candidates who claim they have been victimized by the polls. No matter how well founded, such criticism is usually written off as sour grapes. Louis Harris says: "Most of the criticism of polls stems back to the eyes of the beholder. It's like the ancient Per-

sian messengers. If you bring bad tidings, they want your head."
No winners go around griping about the polls, and it is only the
winners who have the power, by virtue of their election, to
change things.

Ordinarily, disgruntled politicians only make resentful comments to the press. Once in a while, however, they do more. In
1966 Zolton Ferency, the Democratic candidate for governor of
Michigan, took a pollster to court.

Market Opinion Research did the polling for the *Detroit News*.
Ferency had the unenviable task of running against George Romney, who at the time was considered the front-runner for the 1968
Republican presidential nomination; Romney obviously wanted
to give an impressive demonstration of his vote-getting power.
Nevertheless, Ferency felt he was making headway against the
incumbent, but every time he took one step forward, adverse poll
results headlined on the front page of the *Detroit News* would
knock him two steps back. Romney was given such a great lead
early in the race that no one was taking Ferency seriously. Ferency was particularly disturbed, and rightfully so, by the fact
that the *News* had not only endorsed Romney but was using
Romney's personal pollster, Market Opinion Research!

Usually, if a pollster wants to leak information to the press
which will make his client look good, he has to be friendly with
a reporter or columnist who will slip the information into the
paper. Market Opinion Research did not have to go through
these channels; it had direct access to the front page.

In polling for the newspaper and a candidate at the same time,
the firm was involved in an irreconcilable conflict of interest. It
could either present the newspaper with legitimate information,
put in proper perspective, or it could help promote Romney's
fortunes by making him look as strong as possible. It could not
do both.

Ferency sought an injunction prohibiting publication of the
final Market Opinion Research poll in the *News* unless the size
and composition of the sample, the polling methods, and the
interpretation were revealed. As expected, the defense countered
with the argument that any injunction would be a violation of the
First Amendment guarantee of an unrestrained press. The defense also argued that disclosure of the polling procedures would

cause great harm to the pollster, because they were "trade secrets."

The court denied Ferency's request for an injunction, and did so without reaching the difficult constitutional and trade secret questions. Instead, it ruled that Ferency could not prove that he was harmed by the poll. One must show likelihood of serious harm in order to justify the issuance of an injunction, and the court said that any harm in Ferency's case was purely speculative.

If this particular judge's reasoning was adopted generally, trailing candidates could never compel disclosure of polling methods, because they cannot prove that they would win the election were it not for the adverse poll. Indeed, one wonders whether the judge decided on the basis of the *News* poll that Ferency had no chance to win, ergo there was no harm done by the paper's printing it. By this Alice in Wonderland logic, a poll becomes its own test of accuracy.

If harm must be proven to a certainty, candidates can never resort to the courts for help in a battle against the polls, as it is impossible to prove what would occur but for the existence of the poll. Elections are too complicated; there are too many indirect factors. Nevertheless, the electoral process is so fundamental to our political system that anything which even threatens to disrupt it must be carefully scrutinized. The mere possibility of harm thus is enough to warrant action. Ferency ultimately did lose the election. We cannot ever know whether the adverse polls were a factor, but it is noteworthy that the final *News* poll did underestimate his vote by 10 percent.

Politicians rarely sue pollsters. It is even more rare that they prevail, but it has happened. During the 1964 presidential campaign *Fact* magazine published an article called "1,189 Psychiatrists Say Goldwater Pyschologically Unfit to Be President." The title was ostensibly drawn from the results of a poll of psychiatrists conducted by the magazine.

After the campaign Barry Goldwater successfully sued the publisher of *Fact*, Ralph Ginzburg, for libel; the legitimacy of the poll was one of the key questions in the case. Goldwater's lawyers demonstrated that the questions were slanted against him in both their wording and the context in which they were asked: "Does he seem prone to aggressive behavior and destructiveness?" "Can you offer any explanation of his public temper tantrums?" and "Do you believe that Goldwater is psychologically fit? No or yes?" There were no questions in the survey about Goldwater's opponent, Lyndon Johnson.

Goldwater's lawyers also pointed out that the poll had been conducted by mail. Only about 20 percent of the twelve thousand doctors questioned had bothered to respond. It is quite likely that many of those who did had an anti-Goldwater ax to grind, so the sample was not representative. Moreover, the questions did not make clear whether personal or professional opinions were sought. Finally, in its analysis of the poll *Fact* had emphasized only the information which was unfavorable to Goldwater; they quoted psychiatrists who had supposedly said the senator was "a latent homosexual" and "a mass murderer at heart."

The publisher's lawyer could not very well defend the accuracy of the poll; his only defense was to call all polls into question, so that by comparison Ginzburg's survey would not look so bad. Pollster Burns Roper had testified on behalf of Goldwater that the *Fact* survey was totally lacking in the scientific safeguards necessary to insure reliable results. When Ginzburg's lawyer began cross-examining Roper, he led with the question, "Tell us, with all your scientific safeguards, which candidate did you say was going to win the 1948 election?"

Ginzburg won that little exchange, but it is easy to see why he lost the suit. His poll was an extreme example of how questions can be slanted in order to elicit a particular response, and how such responses can be dressed up to say almost anything.

Though Goldwater himself was successful (he was awarded fifty thousand dollars in damages), his suit demonstrates the difficulty that politicians who have been harmed by polls encounter when they seek relief in the courts. Goldwater prevailed because he was able to prove libel. It is one thing to say that a candidate is psychologically unbalanced, but it is another to say that he is trailing his opponent by 20 percent. Though both statements are harmful, libel is clear only in the former case. In a way, then, Goldwater was lucky that the poll was biased against him in such a flagrantly personal way.

Secondly, even though Goldwater won, it was an after-the-fact victory. For the most part, courts are set up to compensate for harm after it has been inflicted, not to prevent it from happening. Zolton Ferency went to court to prevent further publication of the *Detroit News* poll until there was disclosure of polling methods, yet even he had already been harmed, perhaps irreparably, by the *News* surveys which had been published before he brought suit. There is no practical remedy. In theory, a court could order a new election, but in fact the best that a losing candidate can expect from a suit is money damages as a kind of consolation prize.

Most important, the courts are not really equipped to solve the problem of the effect of polls on elections, because they must work on a case-by-case basis. What is needed is a broad policy which applies in advance to all elections, not just a resolution of differences between one candidate and one pollster. It would be difficult in a judicial proceeding to assemble all the witnesses and gather all the information necessary to formulate such policy.

The courts thus may occasionally handle disputes involving polls and politicians, but if there is going to be any large-scale change in the role which polls play in our political system it will not be brought about by judges.

It should also be noted that an individual politician who decides to criticize a pollster may be taking certain risks. Pollsters are allowed to bring lawsuits as well as defend them. In 1970 pollster Mervin Field recovered a $300,000 judgment in a defamation suit he brought against William Penn Patrick, who had been a candidate for the Republican nomination for lieutenant governor in 1966. During that campaign, Field's poll had reported that Patrick would get only 1 percent of the vote. Patrick responded to that bad news by claiming that one of his opponents, former San Francisco Mayor George Christopher, had "bought and paid for the poll."

Politicians are naturally reluctant to accuse pollsters and their publishers of actual bias, for they know that the newspaper always has the final word. The result in the Field case may make politicians even more tempered in their criticism.

If the courts are not going to provide any sweeping remedies to limit the effect of polls on elections, what of the legislature?

The West German parliament has banned publication of candidate preference polls during the two weeks preceeding election day. The broadcasting of such surveys is similarly prohibited on election day in Britain.

There have been scattered attempts in states and cities in this country to regulate the polls. In early 1968 Mayor R. W. Grady of Rockledge, Florida—population fourteen thousand—proposed that the city council draft an ordinance forbidding national polls from surveying that city's residents. Grady said: "All they're doing is influencing the voter. They're no longer reporting voter opinion. It is a dangerous trend when pollsters can call the shots."

Grady's proposal got nowhere, a fate shared by most of the similar proposals which have been made in other localities. In any case, if legislation is to be effective, it has to apply nationally. The only substantial effort for federal action has been Michigan Congressman Lucien Nedzi's proposed "Truth-in-Polling" bill, first introduced in 1968 and re-introduced several times since then.

There had been sporadic congressional interest in polls, but it never came to anything. In the thirties, after the *Literary Digest* predicted that Alf Landon would beat Franklin Roosevelt, there were calls for a special investigation of pollsters. In 1943 Senator Gerald Nye proposed a bill which would have required pollsters to disclose the size of the sample they used and to preserve all their records for two years, but no action was taken on his proposal.

Aside from an investigation of television ratings in the midsixties, Congress has shown no interest in regulating the pollsters. A closer look at the recently proposed "Truth-in-Polling" bill shows just what might be done and why Congress has not done it.

Nedzi's proposal, like Senator Nye's 1943 bill, calls for disclosure of polling methods, but on a much more comprehensive basis. In addition to sample size, the pollster would have to reveal when the polling was done, what questions were asked, how the polling was done, and what portion of the sample participated and what did not.

Under the Nedzi bill, this information, together with the name of the person who commissioned the survey, would have to be filed with the Library of Congress within three days of the poll's dissemination. The Library of Congress would thus become the repository for all published polls, both syndicated polls and private surveys which are intended to be seen by the public, so that anyone who wished to investigate the legitimacy of a poll could do so.

Two things stimulated Nedzi's concern about the influence of polls on the political process. The first was Lyndon Johnson's incessant references to the polls on Vietnam as if they were an incontrovertible plebiscite on the war. (Nedzi himself was an early opponent of the war.) The second was Nelson Rockefeller's unprecedented attempt in 1968 to win the Republican presidential nomination by influencing the polls through massive advertising. Nedzi was particularly disturbed by the clumsy attempt of the Harris and Gallup firms to reconcile totally contradictory polls just before the Republican convention.

Nedzi was able to get sixty co-sponsors, both Democrats and Republicans, but the Committee on Elections, to which the bill was referred, took no action; there were not even any hearings held on it.

In subsequent sessions, Nedzi re-introduced his bill. Events in each succeeding year made him more convinced that some sort of action was necessary. In 1969 the *New York Daily News* reported that John Lindsay was twenty-one points ahead of Mario Procaccino in the mayoralty race in New York, but Lindsay won by only four. In 1970 Senator Charles Goodell bitterly claimed that a *News* poll which showed him trailing both his opponents cost him the election. Then, in 1971, there was great controversy about the accuracy of a poll for the Philadelphia mayoralty race which turned out to be based on a sample of only 157 people.

Public opinion polls were in the forefront of the 1972 presidential campaign. First the polls gave Edmund Muskie an unrealistic early lead, then an erroneous survey by the California Poll said that McGovern was going to win that state's primary by the largest landslide in history. All these incidents heightened interest in the reliability and impact of polls in general, and in Nedzi's bill in particular.

By 1972 Nedzi had become the chairman of the House Subcommittee on Libraries and Memorials, a body which usually has nothing to do with bills on elections. Nedzi had not been able to get a hearing for his bill before another committee, so he arranged to have it referred to his subcommittee on the slender jurisdictional argument that the bill required filing of information with the Library of Congress. As chairman of this rather obscure subcommittee, he nevertheless was able to orchestrate an impressive hearing, with most of the major pollsters testifying, as well as some prominent politicians who felt they had come out on the short end of particular public opinion surveys.

The testimony of the pollsters illuminated many of the pitfalls of polling and also pointed up some of the difficulties in dealing with them. It provided a revealing glimpse of how some of them think of themselves and their profession. George Gallup, one of the first witnesses, expressed support for the bill, as he had from the time it was originally filed. Gallup emphasized that in his releases to his client newspapers he already includes most all the information required by the bill. Gallup said that the principal problem lay with polls released on television or leaked to the newspapers.

To Gallup, the issue of polls and politics is simply whether the polling methodology conforms to accepted practice. He has said, time and time again, "It is obvious that when a field grows as fast as this one has, you'll have some incompetents." Gallup is no doubt right in saying that there are incompetent pollsters and that at least some of their errors would be exposed by mandatory disclosure, but even if all the charlatans were driven out of business, there would still be reason to worry about the effect of polls on the political process.

The matter of technical competence is really secondary. In some respects, Gallup's testimony in support of the bill seemed to smack a bit of the guildsman who is justifiably concerned about the standards of his profession but who also wants to keep out competitors.

Gallup's chief rival, Louis Harris, disagreed with him, as he often does. Harris said that while he agreed that there were some serious abuses, particularly when candidates leak polls, the Nedzi bill presented a "grave danger" to freedom of information. When pressed, he admitted that he was opposed to any legislation directed at the polling industry. He could offer no specific objections to the Nedzi bill, but still was adamantly against it on the ground that it would set a precedent for more restrictive legislation.

Though Harris strongly disagreed with Gallup's support of the bill, he echoed some of Gallup's self-protective sentiments. "The problem, I think, arises when some polling outfit nobody has ever heard of comes along and is quoted in the newspaper or television or radio, or wherever, and is quoted as the gospel, as though it were the Gallup or Harris poll."

That is an interesting trinity: the gospel, the Gallup poll, and the Harris poll. The only surprising thing about his remark is that he included Gallup in such select company!

The Nedzi hearings drew a good deal of attention in the newspapers, in part because they took place during the 1972 presidential campaign, when the national polls all showed Richard Nixon beating George McGovern by an astonishing two-to-one margin. The polls had become a major issue of the campaign.

Unfortunately much of the press saw the hearings through Gallup's eyes, reducing the question to the good pollsters versus the bad pollsters. This view was typified by a *New York Times* editorial which stated, "While such major organizations as Gallup and Harris have built up a deserved reputation for integrity, there is no doubt that polling is a field with more than its share of fly-by-night 'experts' and downright charlatans."

As the *Times* saw it, any problem presented by the polls could be solved by requiring full disclosure of technical information. "No intelligent judgment about any poll results can be made without knowing how big the sample was, when the questions were asked and exactly what they were, and, most important, who paid for the poll—an independent organization or a partisan interest."

It is ironic that the *Times* took such a simplistic view of the proceedings, for their own pollster, Daniel Yankelovich, was the most eloquent of all the pollsters in testifying about the more subtle problems of polling. Unlike most of his colleagues, he acknowledged that the polls could have an influence on the outcome of elections; in fact, he suggested that the national polls at that time were probably having that effect on the presidential race.

Yankelovich explained that both his surveys and Harris' indicated a massive Democratic defection from McGovern toward Nixon. Yankelovich suggested that such figures could become a self-fulfilling prophecy by making it easier for traditional Democrats to vote against their party. "When you see these polls that are reporting this massive Democratic appeal of Mr. Nixon, it might cause these people to feel, 'Other people are doing it,' which reinsures or reinforces that impulse. I do not know how big an influence that is, but I think it could be an influence, and it disturbs me."

The *Times* to the contrary, the issue of polls is not simply whether the pollsters have reputations for "integrity," for the very fact that Gallup, Harris, and Yankelovich are well respected made their polls all the more influential.

Yankelovich expressed general sympathy for the bill; he had none of Harris' concern that it would be an entering wedge which would lead to undesirable legislation. He did, however, suggest that the Nedzi proposal did not really reach what he regarded as the major source of polling error. The proposed bill was principally concerned with technical matters of sample design and interviewing technique, while Yankelovich was more concerned with the more subjective area of the nuances of question wording.

Yankelovich stated that the reluctance of newspapers to print even the technical information, sample size, and the like made him very pessimistic about the chances that the press would on its own deal better with the more important matter of interpretation. "If there is difficulty in getting media to put in one little box those few very specific kinds of qualifications, the difficulty is magnified many-fold by getting them to put all the possible interpretations, not only the wording of the questions, but some of the things that the answers to that question might mean."

The *Times*, and other newspapers, failed for the most part to pick up the two points which Yankelovich made: first, that polls can have an impact on elections, and, second, that the pressures of what he characterized as "horse-race journalism" lead to superficial and misleading interpretations of the polls. These are critical issues, but they were lost on the press.

Yankelovich was not alone in asserting that the major source of error is not statistical but interpretative. In his statement

Burns Roper warned: "I am quite opposed to the inclusion of a sampling error statement in the report of the poll results. I think the implication of a sampling error statement is that it is the measure of total error in a survey."

Roper noted that sampling error could throw a poll off by several points either way, but that really is insignificant compared to other sources of error. "Question error—the bias or loading in question wording, or the error that results from the context in which the unbiased question may occur—can cause errors of ten, thirty, or even more percentage points."

Contrary to what might have been expected, the Nedzi hearings were not simply a showcase for the pollsters to congratulate one another on their insight and importance. Yankelovich and Roper raised some sophisticated issues. Even Gallup, the patriarch of the polling family, expressed the need for some remedial action.

Not all the pollsters, of course, were as helpful. Mervin Field, the proprietor of the California Poll, went through lengthy contortions to prove that his pre-primary survey which had shown McGovern beating Humphrey by twenty points really was accurate, in spite of the fact that the actual margin was only five. Unfortunately, Fred Currier of Market Opinion Research was not cross-examined about the events which led to the lawsuit brought by Zolton Ferency, mentioned earlier.

A whole range of pollsters, from Dr. Angus Campbell, the director of the Institute of Social Research at the University of Michigan, to Jimmy the Greek, the Las Vegas oddsmaker, did provide useful information on their polling methods, as well as their views on the impact of polls on politics.

There were also politicians, a parade of walking wounded from the battle of the polls. Zolton Ferency testified about his run-in with Market Opinion Research. Former New York Senator Charles Goodell said that adverse and inaccurate polls in the *New York Daily News* had killed his campaign. Congressman William Green, who had privately expressed great bitterness about the crushing effect a newspaper poll had on his campaign for mayor of Philadelphia, was much more moderate in his public remarks.

When the hearings were concluded, the committee had gathered a remarkable amount of information reflecting a wide variety of experiences and points of view. The congressmen on the committee showed varying degrees of knowledge about the polls, but a reading of the transcript shows that each of them grew more sophisticated as the hearings went along. That the hearings were able to attract the major pollsters is a credit to Nedzi's perserverance. That the questioning was for the most part intelligent and illuminating reflects the unusually high quality of the staff work by counsel Jack Boos and others.

Why then did nothing happen? For all the preparation, and in spite of the national attention the hearings received, Nedzi's bill got nowhere.

There are several explanations. First, the bill was concerned with the disclosure of technical information about the polling process, but the testimony of the pollsters showed that there was little agreement as to what constitutes good methodology. Gallup and Harris were critical of telephone interviewing, for example, while Yankelovich and Sindlinger maintained that it could be as reliable as personal interviewing in some situations, and perhaps more so.

If the pollsters themselves cannot concur on what is proper technique, it is less clear that disclosure can really help people evaluate polls. To compound the problem, some of the pollsters said that their practices were trade secrets, disclosure of which would mean they would lose an edge to their less sophisticated competitors.

Several members of the committee were also concerned with possible infringement of First Amendment guarantees of a free press. Some seemed to assume that any restriction on the publication of polls is necessarily unconstitutional. The requirements of disclosure under the Nedzi bill, however, impose only a slight burden, particularly in contrast to the potentially great harm that polls can do to the electoral process. Although the matter has never been before a court, it seems likely that legislation such as that proposed by Nedzi would withstand any constitutional challenge.

Perhaps the most significant reason the bill died quietly is that Congress is at best apathetic about the influence of polls on politics. Nedzi was able to drum up interest within his committee, but few outside it were aware of the bill or the hearings. When questioned in 1975, Majority Leader Tip O'Neill and Judiciary Committee members Peter Rodino and Charles Wiggins all said that they had never heard of Nedzi's attempts to sponsor legislation on polls.

This apathy should not be surprising. Those who are in Congress have survived the polls, perhaps even benefited from them. If there were a shadow government made up of people like Charles Goodell, Zolton Ferency, and Mario Procaccino, their first order of business would be to pass a statute on polls. But Congress belongs to the winners, and there is no strong constituency for change in the polling status quo. For that reason it seems unlikely that the Nedzi bill, or legislation like it, will stand any better chance in the future.

That is not to say, however, that Nedzi's efforts were futile. He introduced his bill primarily to raise awareness of the mechanics of polling and the impact of polls on the political process. The hearings unfortunately left no lasting impression on the congressional leadership. Though they did attract notice in the press, that too was transitory. Yet the fact that legislation was proposed and thorough hearings were held was not lost on the pollsters themselves. Even though the prospect for passage of the Nedzi bill, or some other polling legislation, may be remote, the mere possibility may induce some of the more responsible pollsters to take a second look at the implications of what they do.

There have been attempts in several states to regulate polling practices to some degree. Legislation was proposed in California, for example, which would have required pollsters to register and be licensed under a system establishing standards of professional competence. Like the Nedzi proposal, the California bill never got out of committee.

The New York legislature rejected a bill which, among other things, would have required the disclosure of the names of all the people who are polled. That provision would have made it even more difficult for pollsters to complete their interviews, as people would justifiably fear that their remarks would become public record. New York, however, did enact a more modest statute which requires any candidate who discloses the results of a poll to reveal some of the basic methodology on which it was based. The statute is apparently the only regulatory law in the nation governing polls; although it is not stringent, it has already proven useful.

In 1974 Irwin Harrison, then of Decision Research, was polling the New York gubernatorial race for some local newspapers. His surveys showed that Democrat Hugh Carey was running very strong, but his client newspapers had gotten wind of a poll which showed Republican Malcolm Wilson ahead, and they wanted to be sure their pollster was right. Because of the disclosure requirements, Harrison was able to learn that the poll was for a county of 700,000 people and had originally been based on a sample of 300. The figures showing Wilson ahead, however, were based on re-interviews with only 158 of the original sample. It was clear to Harrison that the sample was both too small to give statistically reliable information and probably biased in favor of Republicans, who were somewhat easier to find at home.

While the New York legislation is most welcome, it is doubtful that it represents the wave of the future. In most state legislatures and in Congress there simply is not a constituency which is concerned enough about the impact of polls to get regulatory legislation enacted.

If the legislatures are not inclined to act and if the courts are unable to deal with polling problems on any but a case by case basis, is there any reason to believe that the pollsters will regulate themselves?

There are three professional organizations for pollsters. The American Association for Public Opinion Research, founded in 1940, is the oldest and largest. Its members are individuals, as opposed to firms, the great majority of whom do polling for private business, government, advertising agencies, and the like. Only a small portion of them are engaged in publishing public opinion polls, but most of the best-known pollsters, including Gallup, Harris, and Yankelovich, are members.

AAPOR seems to have two basic purposes. One is to act as a clearinghouse for technical information. It publishes the *Public Opinion Quarterly*, a scholarly journal on polling, and it also hosts regular conferences at which ideas and experiences are exchanged. AAPOR, like most other trade associations, exists to enhance the image of the calling. Indeed, the underlying reason for its creation was to bootstrap the business of polling into a profession.

AAPOR has only recently become actively involved in problems of ethical responsibility, and even now its involvement is quite tentative. Dr. Angus Campbell, the director of the Institute of Social Research at the University of Michigan, recalls that when the question of mandatory professional standards first came up at AAPOR meetings in the late 1940's, most of the membership was strongly opposed. "At that time the establishment figures of the polling business were flatly opposed to standards of any kind and quite frankly defended a 'let the buyer beware' position."

That attitude has changed somewhat over the years; witness Gallup's support for the Nedzi bill. In 1950 AAPOR did establish an ethics code, and in 1968 it promulgated standards for disclosure of information when polls are intended to be disseminated to the public. Under the ethical code, a member is supposed to retract any poll released in a misleading manner. The disclosure standards are intended to serve the same function as the Nedzi bill, that is, to provide sufficient information so that readers can make an independent judgment about the technical quality of a poll.

There is a standards committee which is supposed to investigate complaints of violation of the professional code, but the committee has been reluctant to use the limited power it has. It does conduct investigations, but it usually makes only factual reports; it does not pronounce any judgments as to whether the conduct in question was unethical. The committee's factual findings are sent to the members, but they are not released to the general public. Though undoubtedly there have been circumstances to warrant it, no member of AAPOR has ever been censured or expelled.

There are some legitimate reasons why AAPOR has been unable to deal forcefully with polling abuses. One of the most flagrant misuses of polls is leaking of misleading information, but often it is hard to track down who took the poll and what it actually said. Then too, AAPOR has no control over the actions of non-members. As Gallup points out, "We are a small group, 5 percent of the total, who regulate ourselves and follow our own rules, but we have no way to get these other people into this organization."

The underlying reason why AAPOR has been so hesitant to pass judgment on its members' work is that pollsters are not really sure what constitutes good polling methodology. Sidney Hollander, who was president of AAPOR at the time of the Nedzi hearings, testified before the committee, "Research comes in so many sizes and shapes, and with so many different goals and purposes, that there can be no universal standard for what constitutes a good poll."

The National Council on Public Polls was founded in 1968, in large part because pollsters recognized that AAPOR was not equipped to deal with the growing problem of abuse of polls. NCPP is a much smaller organization; its members are polling firms who are for the most part concerned with published, as opposed to private, polls.

Its focus, much more narrow than AAPOR's, is on educating the public in general and the press in particular about reporting and reading polls. Operating with a very modest budget, it has sent newsletters to editors and broadcasters suggesting the establishment of more sophisticated standards in media treatment of polls. Its disclosure standards are virtually identical to those of AAPOR, and there is cooperation between the two groups.

The Nedzi committee also heard Robert Bower, who was then the president of NCPP. He was candid in admitting that his group has been able to do little to prevent the publication of erroneous and misleading polls. "We are trying to do more," he told them, "but I think 'puny' is a good word for our efforts so far."

Brower echoed Hollander's explanation of the difficulty in applying professional standards. "Our profession is a combination of science and art, and lends itself to tremendous differences of opinion as to what size of a poll is adequate or inadequate, what is a good question and what is a bad question, and so forth and so on. So there are very arguable points and it is very difficult to say yes, we have these standards, these specific standards, and someone has not met these standards."

Brower's frankness is commendable, of course, but the clear implication of his statement is that polling methods are so crude, or elusive, if you prefer, that it is impossible to say that something is a bad poll. Because there can be no set standards, there can be no meaningful sanctions for abuse.

George Gallup agrees that AAPOR and NCPP as presently constituted are not really equipped to get to the heart of the problem. "We considered putting teeth in some of these regulations and actually kicking out members who didn't conform, but then, by God, you run into the possibility of lawsuits and everything else in the world."

It thus seems unlikely that the pollsters collectively will do more than they are doing now. Brower of the NCPP wishes that newspaper coverage of polls could be monitored, so that a pollster would know how his data was being used, but the NCPP does not have anywhere near the necessary funds for such an undertaking.

In late 1975 the National Association of Private Pollsters was formed by Bob Teeter, Pat Caddell, and several others in the hope of making the Federal Election Commission responsive to the problems of political pollsters and their clients. At present, the bipartisan group's membership is small and its objectives are limited, but its founders hope that when the organization grows, it will promulgate professional standards.

Whatever the pollsters do in the way of self-restraint will probably be on an individual basis. Daniel Yankelovich and Peter Hart, two of the most intelligent and responsible pollsters, have had some success in educating their clients, the *New York Times* and the *Washington Post* respectively, to play down the horse-race aspects of polling and to acknowledge the interpretive side of the process.

The question of self-restraint involves not only how polls are presented but whether polls on certain issues are presented at all. Before the Nedzi committee Louis Harris said, "We now exercise restraint on any law case." He admitted that his interviewers had sometimes asked questions about the guilt of a Sirhan Sirhan or a James Earl Ray, but that the results were never released publicly. Publication, he said, could interfere with the court process by influencing jurors to vote for the popular verdict.

Yet even though Harris acknowledged the danger of such polls, he said that he would be strongly opposed to any legislation

which would prohibit such publication. As with the Nedzi proposal, he claimed to be opposed not to the substance of such a law,
but to the creation of a precedent for still further regulation. The
decision as to what ought to be published and what ought not to
be must, Harris was saying, rest with the pollsters themselves.

The pollsters cannot be trusted, however, to have the discipline to say that they will not poll on a sensitive subject. During
the most important legal case in the nation's history, the Watergate scandals, the pollsters privately wrung their hands about
whether it was proper for them to keep a running score on what
percentage of the population thought Richard Nixon was guilty
of a criminal offense, yet whatever their misgivings, they could
not resist temptation and published anyway. The questions were
often badly worded and the results misleading, but the polls on
impeachment had a profound effect on our history in 1973 and
1974.

If government seems unlikely to police the polls, either
through the courts or through the legislature, and if the pollsters
do not seem able to regulate themselves, is there any other institution or body which can deal with the problem? At the Nedzi
hearings, AAPOR president Sidney Hollander said, "In the long
run we do believe that the solution will be greater sophistication
of the media and, through the media, their audiences."

In some respects, hoping that the press will cure the abuses of
polling is as unrealistic as expecting that prostitution will die
because of lack of customer interest, for it has been the press
which has been responsible for the most flagrant and damaging
misuse of polls. NCPP president Robert Bower told the Nedzi
committee, "I think at present there is just as much external
pressure on the pollster to flout the standards of his profession
as there is for him to uphold them."

Much of that pressure comes from newspapers. Archibald
Crossley stopped publishing his surveys "because editors want
you to stick your neck out and predict." Before Harris began to
syndicate his work, he too questioned the emphasis the press
gives the polls. "Part of the problem for public polls is that
newspaper editors are far more interested in the game of political
Russian roulette (is the poll right, and, if so, within how many

percentage points?) than in a comparison of the 'whys' behind the voter's decision."

Daniel Yankelovich similarly complains that "there is a real disjuncture between what we pollsters regard as significant and what the press imposes. The horse-race aspect of polling is the least interesting, least meaningful aspect of the whole process, but it's what the press wants. That's the bugaboo of our profession."

There are some signs, however, that a few newspapers are becoming a bit more careful about the polls they print. A number of papers, such as the *New York Times*, the *Washington Post*, and even the *Boston Globe* (which is one of the more poll-happy papers in the country), have taken to providing a brief summary of technical information with each major poll they print. The information they provide is parallel in many respects to that required under the Nedzi bill or by the AAPOR and NCPP standards of disclosure. The reader thus knows how many people were in the sample or when the interviewing was conducted.

Often this sort of information is printed in a little box which also may contain some warning that the results are not to be construed as predictions. The increasingly frequent appearance of this box is welcome in that it does give the sophisticated poll reader some basic information with which to make tentative judgments about the technical quality and limitations of the poll.

The box, however, also has some drawbacks. First, it is in some respects like the surgeon general's warning which must appear on all cigarette packages and advertising. Although it suggests that there may be dangerous to use the product, the context in which the warning appears tends to nullify its affect. A billboard of a man happily smoking a cigarette while floating in a sylvan pond is supposed to sell cigarettes in spite of the somber warning that they have been found to be hazardous to your health. Likewise the box which appears next to the newspaper poll may suggest that the reader not make too much of the percentage standings, while the headline screams that one candidate is leading another by five points.

If anything, the box in the newspaper tends to strengthen the impact of the poll, as if it were being certified Grade A. The statement that there is a range of sampling error of several points either way leads most people to believe that the poll as a whole must be accurate to that degree. Of course, sampling is but one source of error, and a relatively small one at that.

The second problem with the box is that it makes the newspapers feel that they have done all they have to in order to be responsible in their treatment of polls, even though presenting the technical information is only a modest first step in that endeavor. Yankelovich cautions: "All that little box can tell you is that the poll is scientific in the limited sense that if a reputable pollster reports that 72 percent of the people say thus and so, then you can be pretty sure that's what they said. But what those people meant, as opposed to what they said, is the crucial issue, and there is no way that a little box talking about the sampling method is going to shed any light on that." If anything, the presence of the box lulls people into thinking that the figures can be taken at face value, when in fact they may be extremely misleading.

The emphasis on the box directs people to the technical questions of polling, when they should be questioning the subtleties of interpretation. Peter Hart shakes his head about a columnist who once earnestly told him that he had memorized "the seven questions to ask to see if a poll is accurate." The questions all involved sample size, method of interviewing, and the like. Hart says: "There is no way any seven questions or any box is going to help you understand the substance of a poll. A meaningless question will still produce meaningless results even if it is asked in a scientific way."

Unfortunately, this is lost on many members of the press. Yankelovich puts his finger on the problem. "What makes the matter complicated is that the solution is extremely unobvious. You are put in a position of saying that people are misled by the facts, by hard, demonstrable data." The black and white statistics of the polls seem to be clear and conclusive, particularly when they carry with them a box certifying that they are scientifically accurate, yet if they are not set in proper perspective they can be dangerously misleading.

What is needed is a change in orientation away from straight numbers and toward more analysis. Yankelovich believes that too many of his colleagues approach every polling issue as if it were an election in which people can only vote for or against. This two-dimensional approach removes the texture and depth from public opinion. He suggests that newspaper polls would be greatly improved if they got away from using just one indicator as a test of public opinion on a complex issue. "You can find out what people think, if you talk with them long enough, so you can still report on public attitudes, but instead of presenting just one item, you show the picture in its entirety."

Though there were numerous polls on the war in Vietnam, the one question which dominated the headlines and was taken as the definitive test of public opinion was "Do you have confidence in the president's conduct of the war?" The question was, however, less an indicator of attitudes on foreign policy than it was a test of depth of loyalty to the government no matter what policy it pursues. Similarly, the question "Do you think President Nixon should be impeached?" seems direct and to the point, but it produced extremely misleading results because many Americans did not really understand the impeachment process.

Peter Hart believes that, together with improvement in interpretation, there must be more imaginative questions asked. To him, the agree/disagree question is the bane of polling, for it forces complex opinions into artificially narrow and misleading categories.

Hart sees a variety of directions which polls could take. "Why do we never try to understand what the public reaction is on issues, and then relate it back to candidates? We would understand better that such and such a candidate is going to run into a whole hell of a lot of trouble because at some stage his ideas are going to run into direct conflict with those of the electorate. It's on this kind of thing that the polls are falling short."

Hart is a great believer in the potential of newspaper polls to inform and educate, but he admits that they have not fulfilled their promise. On election polls most pollsters play down the undecided voters, as if they somehow do not really count. In fact, in many cases these voters hold the balance of power in an election. An intensive study of that particular group may unlock the basic dynamics of the campaign. That a voter says he is undecided carries at least as much meaning as a statement of support for one candidate or another. There may be real indecision there, or perhaps a hidden vote for a controversial candidate, or deep alienation from the political system.

Newspapers play up trial heats between various candidates months, even years, before an election, but these polls tell us little of a candidate's strength and weaknesses. It would be far more revealing for the pollsters to report from time to time how well known the potential candidates are. The few pollsters who do this do it in a superficial way, testing only name familiarity. A typical Harris poll, for example, will ask people to indicate which politicians they have heard of out of a list of twenty or so. Harris takes people at their word. The people are not tested to see whether they really have heard of a particular politician, or even, if they have, just what it is they know.

Peter Hart is one exception. He has tried to develop more revealing tests of substantive recognition of candidates. Hart's surveys indicating that relatively few people knew anything substantive about Senator Lloyd Bentsen but the fact that in mid-1975 Bentsen was trailing President Ford in the Gallup and Harris pairings by eight or ten points in an entirely different light. That Bentsen was able to do so well, relatively speaking, said something about him and perhaps even more about Ford.

It is this kind of probing—the use of many questions to define the outlines of opinion and to explore ignorance and the depth of feelings—which can lead to more meaningful polling. But just as the statistics must be backed up by the little box which explains how they were derived, so must any interpretation be accompanied by enough information so that the reader himself can judge whether the analysis is sound.

Louis Harris prides himself as being a public opinion analyst; he emphasizes that all his columns are based on ten or more questions, the answers to which he analyzes in order to come up with the mood of the country. But Harris' columns are based entirely on conclusions. He does not present the evidence, the actual questions asked—he says that would take up too much space—so there is no way that the reader can form an independent judgment. It is as if one were trying to learn literature simply by reading reviews: to attempt it, one must have an unquestioning faith in the wisdom of the reviewer.

Harris complains that he has limited space and that editors

have total discretion to cut whatever they please out of his releases. Perhaps this is so, but both Hart and Yankelovich have
had far better luck with their newspapers, which reflects well on
both them and their clients.

When Yankelovich first began dickering about doing work for
the *New York Times*, editor Abe Rosenthal admitted that he was
generally anti-poll. Yankelovich replied that from his point of
view the interesting part of polling was reporting on the underlying political fabric—how issues and personalities interrelate,
what moves the electorate, and so on. He also cautioned that this
would take much more space than goes to the traditional poll.

Yankelovich was hired, and his newspaper work is considerably higher in quality than that of Harris or Gallup, largely because the *Times* has given him free rein to do what he thinks is
important. Much of his work has been innovative and informa-
tive. During the 1972 primary campaign, for instance, he in-
stituted the practice of surveying people not just before elections
but afterwards as well. He interviewed people after they left the
voting booth to discover not just for whom they voted but why.
His reports added much depth to the *Times'* post-election analysis.

Hart has had similar success with the *Washington Post* and the
other newspapers for whom he has polled. He too has gone
beyond the simple horse-race coverage of election campaigns.
For the 1972 Pennsylvania primary he asked not just whom people would vote for, but whom they would not. He discovered that
50 percent of the electorate disliked George Wallace so intensely
that they said they could not vote for him under any circumstances. Hart says, "That's where polling can really provide a
different dimension on an election. You could study all the returns you want to and never come up with that very essential
piece of information."

That Hart and Yankelovich have been able independently to
break new ground offers some hope that newspapers will move
away from the simplistic and misleading treatment they have
given the polls. Harris claims emphatically that he is at the mercy
of editors, but in truth he probably would have some muscle if
he chose to use it.

No doubt there will continue to be strong pressure to produce polls which make flashy headlines, but if the pollsters are strong enough to dictate the terms by which their work is used, and the well-known ones are, these pressures can be overcome. Perhaps the orientation of the night editors and the headline writers will never change, but responsibility should not rest solely with them. Political writers and columnists, particularly the most respected ones, such as David Broder, Joseph Kraft, and James Reston, must see to it not only that they avoid glib and misleading use of polls themselves, but that the newspapers which carry their work become more sophisticated and responsible.

In 1960 Henry Cabot Lodge, then the Republican vice-presidential candidate, observed that public opinion polls were a passing fancy, soon to disappear from the American political scene. "In the future, people are going to look back on polls as one of the hallucinations which the American people have been subject to. I don't think the polls are here to stay."

Americans have been prone to mass hallucinations before, but usually we quickly snapped out of them. Within several years of the hangings of the "witches" of Salem, our Puritan forebears confessed that it was their own hysteria which had led them astray.

Apparently we are more vulnerable today, for despite Lodge's forecast we have remained completely bewitched by the polls, and there is no sign that the spell will break soon.

Following this chapter, there is a short postscript on how to read polls. These rules will protect an individual from delusions and deceptions, but they cannot solve the problem of the dangerous influence of public opinion polls on our political and social institutions.

This influence will not be eradicated until the press and politicians recognize both that polling has built-in limitations and that public opinion is so complex and elusive that it can never be completely knowable. Polls have become entrenched in part because we have been dazzled by any appearance of science, but more because, living in such rapidly changing, disorienting times, we are desperate for any clue which might tell us what we really think and feel.

The pollsters have offered us what they claim is a mirror, a remarkable construction into which we can gaze to see who we are. The image which is reflected, however, is at best blurred, and the slightest tilt can distort it completely. Public opinion is so fundamental to a democratic society that we must understand its full dimensions, its shape and texture. The polls offer only the trappings of scientific accuracy and the illusion of popular choice. That is not enough.

Postscript
How to Read the Polls

There is an old Maine recipe for cooking coot, a tough and gamy duck. Many people consider coot inedible, but old down-Easters are happy to tell visitors how it can be prepared. After cleaning the bird, drop it in a large pot of boiling water. Add plenty of sliced onions and carrots. Season to taste, then add an old leather shoe. Simmer the concoction for two full days, replacing the liquid as necessary. Then remove from the heat, throw away the coot, and eat the shoe.

A person who wants to understand public opinion might do well to follow the same sort of directions. Read the polls, eavesdrop on conversations in bus stations and coffee shops, scan a variety of newspapers, put a finger in the wind, then forget about the polls.

There are, however, people who claim to eat coot and like it, if it is carefully prepared. By the same token, people can gain some nourishment from the polls, if they are selective about what they digest. Healthy skepticism is a necessity in reading polls, not a luxury. What follows is a series of questions which a reader of polls must keep in mind in order to understand what they mean and whether they can be trusted.

1. Are the numbers right? Americans are beguiled by numbers. We may complain about being digits, not names, in a computerized society, but baseball, celebrated as our national pastime, is less an athletic contest than it is a device for producing statistics like batting averages and runs batted in. Football crowds shout, "We're number one!" The Dow Jones Index measures the pulse of the nation's business in points and fractions. Millions of Americans play the numbers daily. We buy pocket calculators and digital clocks. Small wonder, then, that we are mesmerized by the crisp numbers of the polls.

The careful poll reader must ask two questions: are the numbers right, and what do they mean? As to the first, the reader must know how large the sample was and how it was constructed. For a national poll, a sample of about fifteen hundred should, if fairly drawn, produce results which are within several points one way or the other of what would be obtained if everyone in the country were polled. For a good-sized state, it takes a sample of at least five hundred people to produce the same degree of accuracy.

The typical range of chance error may seem small, but in many elections it can be decisive. As Richard Wirthlin, Ronald Reagan's pollster, has observed: "There seems to be a pseudo-concreteness in the numbers: Reagan 44 percent, Ford 40 percent. But we just don't have that degree of accuracy." When reading a pre-election poll, it is a good habit to juggle the figures in your head to remind yourself how important the sampling error may be. To use Wirthlin's example, a survey showing Reagan ahead 44 to 40 may make it seem that he has a clear lead. Assuming that the poll is based on the typical twelve to fifteen hundred interviews, it is possible that he could be leading as much as 48 to 36, or, just as likely, be behind Ford 40 to 44. All those results are within the bounds of normal chance error.

This chance sampling error is simply a matter of luck. Sometimes the sample will be perfectly representative of the whole population; at other times it will be fairly far off. The larger the sample, the smaller the range of likely error, but diminishing returns soon set in. Most published polls do state the sample size, but many erroneously state that the range of sampling error is the measure of possible error for the entire poll. This, however, is only one possible imperfection; there are many others which can be more significant.

The newspapers rarely tell very much about how the sample was constructed, other than to say it was "scientific." Here the reader must place his trust with in pollster, for he has no way of knowing just how scientific the process was. Most of the better-known pollsters do use techniques designed to give everyone an equal chance of being selected in a survey, but invariably compromises are made for economic reasons, and each of these compromises can potentially throw off the results. Pollsters, for example, are finding that an increasing number of people simply refuse to talk to them. These people are not represented in the polls, and their views may be somewhat different from those who do talk to pollsters.

Beware of any poll which reports percentages down to decimal points. In the past the *New York Daily News* poll has done this; the A. C. Nielsen Company similarly rates a television program's share of the audience down to tenths of a percentage point. It may seem very precise to state that 52.7 percent of the people support a particular candidate, but unless the sample is based on tens of thousands of interviews, the chance sampling error is too large to allow that degree of precision. A pollster who makes that kind of mistake is likely to have made others which may not be so conspicuous. The decimal points should raise a red flag warning that the entire poll is suspect.

While the major pollsters rarely make that particular error, they often make another by presenting percentages for subgroups, such as men and women, people over sixty, and the like, without indicating that the range of chance sampling error for the groups is much higher than it is for the whole.

A typical Gallup or Harris election poll will break down the figures by geographic area, and usually it will appear that one candidate is running stronger in some areas than he is nationwide. While this may in fact be the case, it may also be the product of chance error. Unless the gap is very large—for example, the Democrat leads in the East by ten points or more and is even in all the other areas—you cannot be sure that the differences are significant.

Even the best pollsters are not infallible; they all have used faulty samples at one time or another. But, contrary to popular belief, sampling is their strong suit. It is the one part of the process which is scientific and permits measurement of error. The newspapers usually do provide the sample size and range of chance error (though they often mislabel it). Polling method is important, but it is not the real source of error in polls. Do not let your attention be distracted by asking how fifteen hundred people can possibly represent one hundred and forty million adult Americans, for that is not the real issue.

2. *Have the numbers been adjusted?* Most of the polls we read in the newspapers are not simple tallies of how many people said yes and how many said no. In most cases, the pollster has adjusted the raw figures in one way or another, but unfortunately that is rarely explained.

Election polls, for example, are often based on interviews with "likely voters." Each pollster has his own method for determining who the likely voters are; commonly it involves a series of questions about past voting behavior and interest in the upcoming election. In the end, however, the pollster must make up his own mind which of the interviews to count in his survey and which to throw out. He must also make a projection of voter turnout on election day. Both of these functions involve a lot more personal judgment than science.

This is particularly true for primaries. Democratic poll-taker Bill Hamilton warns: "If there is one thing people ought to look for, it is whether the pollster screened out those people who don't vote in presidential primaries. You can have a candidate who is greatly preferred by the voters at large, but if his opponent has a better organization, the opponent is going to get two or three times the percentage of the vote indicated by the poll."

Unfortunately, few newspaper polls ever spell out how this screening process is done, so the reader has no way of deciding whether the pollster was correct. One can do no more than bear in mind that adjustments have been made and that they involve a large dose of subjective judgment.

Beware also of any poll which shows one candidate leading another by 60 percent to 40 or one policy approved by 55 percent and disapproved by 45 percent. If the numbers for and against add up to 100 percent, then something has been left out: the undecideds. In elections, those who have not made up their minds may be the decisive factor; on issues, the number of people with no opinion may be the most interesting statistic, for it reveals both apathy and ignorance.

Pollsters are instinctively hostile to the idea of people without opinions. They go to great lengths to get people off the fence, both in the interviewing process, by forcing people to state a preference even if they have not really made up their minds, and in reporting the results, when people who are undecided are sometimes thrown out of the sample.

During the 1972 presidential campaign, for example, George Gallup reported, "The Democratic Party currently holds a marginal lead over the GOP, 53 to 47, as the party voters believe can better handle the problem they consider to be most important." In a year when there was great disenchantment with the candidates in particular and politics in general, it was preposterous to think that everybody preferred one party or the other.

Indeed, that was not what people had told Gallup's interviewers. Only 34 percent had thought the Democrats were more competent, compared to 28 percent who favored the Republicans. The largest group, 38 percent, either said that there was no difference between the two parties or did not express any opinion. That figure may well have reflected the alienation and apathy in the country. Gallup, however, simply discarded it, arbitrarily allocating half of the group to the Democrats and half to the Republicans!

In most cases the pollsters probably under-report the percentage of people who really are undecided or have no opinion. People who are wavering are lumped together with those who are firmly committed. A poll which purports to show that everybody has taken a stand one way or the other is so out of line with the way people think that it must be ignored entirely.

3. How was the poll conducted? Pollsters bicker among themselves as to whether personal interviewing or interviewing by telephone is more reliable. The older pollsters remember that one reason the *Literary Digest* predicted that Alf Landon would beat Franklin Roosevelt in 1936 was that a major part of its sample was drawn from telephone books, which during the Depression meant that the survey was heavily biased in favor of Republicans. Today, however, that is far less true, and given the problems which personal interviewers have getting into some houses, the telephone offers some advantages.

Both methods have their strengths and weaknesses, and if the poll was conducted by a professional firm, it should not make any significant difference which was used. Amateur polls, those taken by politicians too poorly financed to hire a professional, are often taken by phone because it is cheaper; in such cases, the use of the telephone may be a clue that the poll was a shoestring production.

Postcard polls are never to be trusted. Most congressmen exploit their free franking privilege by sending periodic questionnaires to their constituents. It is rare that more than 10 percent bother to reply, and their responses usually are not representative of the whole. Indeed the fact that they communicated with their congressman marks them as being different from most people.

Perhaps some congressmen actually think they are polling public opinion, but the more sophisticated among them know that the results are meaningless. The process of asking the voters for their opinions, however, can help create an impression of a concerned and responsive representative in Congress.

There are newspapers, however, which print these polls as if they were scientific surveys. The *Boston Globe*, for instance, gives them as much play as the syndicated polls which it sponsors. It once reported that a poll by Louise Day Hicks, then a congresswoman from Boston, showed that 80 percent of the people in her district were against busing. It was not until the fourth paragraph of the story that it was mentioned that 200,000 questionnaires were sent out and 23,000 responses received. There was no suggestion in the article that this made the results suspect. Indeed the *Globe* quoted Hicks as saying that the response was much higher than that received by other members of Congress and that it "proved decisively that the 'silent majority' does care about the country and its problems, and, given the opportunity, will speak out on the issues"!

It is easy, of course, to poke fun at Hicks' logic—if there was a "silent majority" it was the 87 percent who did not respond to her poll—but the real criticism should be directed at the *Globe*, and papers like it, which print this sort of nonsense.

4. Is it an election poll? Election polls are unique. Unlike all other surveys, they present a clear choice, one that the respondent will shortly have to make as a voter. Regardless of any misgivings, everyone must finally vote one way or the other, or not vote at all. Especially when the election is soon, it is much easier to learn people's true preferences about candidates than their opinions on complex issues.

Election polls are also unique in that the published pollsters put far more effort into them than into any other surveys. They never can be proven wrong on a poll on foreign policy, but they can be embarrassed when an election comes out differently from their polls. That the pollsters have had some success in calling presidential elections (and indeed they have had their share of blunders) thus does not mean that their other work is of equal quality.

The national trial heats between various possible presidential candidates which both Gallup and Harris run regularly tell us little about a candidate's real potential. Respondents are asked to state their preference "if the election were held today," but that is an assumption contrary to fact. Candidates have not geared up their efforts, and people have not given any real thought to the election, so "preferences" are little more than expressions of familiarity. The better-known candidates are often given early leads which turn out to be wholly unrealistic.

The pollsters would perform a much more useful service if, instead of running these meaningless trial heats, they explored the question of how well known the candidates are, testing not just whether people recognize their names but how much they actually know about them. The extent of political ignorance can be amazing. Just two weeks before the 1972 election, the New Jersey Poll reported that only 5 percent of the electorate could identify Paul Krebs as the Democratic candidate for the United States Senate. Republican Clifford Case, running for a fourth term, was correctly identified by fewer than one out of five people!

In spite of this widespread ignorance, the pollsters keep churning out trial heat surveys when it would be much more useful to know what impressions of the candidates the voters do have, and where people stand on the decisive issues.

The pollsters have never been particularly imaginative about pre-election polling. Until March 1972 Louis Harris did not conduct any two-way pairings for the presidential election; he simply assumed that George Wallace would be a third-party candidate. Harris later admitted his shortsightedness, but throughout 1975 he was guilty of the same sin, only this time he assumed that Wallace would not run as an independent, and consequently just ran President Ford against various Democrats. A Wallace candidacy would undoubtedly have a profound effect on major-party strength, but the pollsters gave us no clue as to just what that effect might be.

A simple rule of thumb for reading primary polls is: don't. You can be absolutely sure that the election will come out differently from the polls. The earlier the poll was taken, the greater the difference may be, but even a last-minute primary poll is a chancy thing. Voter turnout is low; in many states people can cross over to vote in the primary for another party; and voters often act on whim or impulse, knowing that the choice they are making is not really final.

Consider a series of *Boston Globe* surveys for the 1972 presidential primary. After the election, the paper congratulated itself, calling its poll "uncannily accurate."

	February 13	April 9	April 20	Election Results
McGovern	11	38	43	51
Muskie	46	27	19	22

It is true that the final poll did correctly indicate that McGovern would win easily, but what of those other surveys? Can they be said to mean anything at all? Exactly two weeks before the election, the *Globe* had McGovern ahead by only eleven points, but he ended up winning by almost three times that number. Pollsters explain such a sequence as evidence of the "volatility" of primaries. What it really shows is that a well known candidate will hold an apparent lead until the campaign really begins; then it is anyone's race. In any event, the early surveys were no guide to the outcome of the election.

The most interesting statistic in election surveys is the one which is given the least attention, the percentage of people who are undecided. Unfortunately, the pollsters tend to minimize the number of people who are undecided, and tell us little about their characteristics.

For example, in September 1975 Gallup reported that Republicans strongly favored the nomination of Gerald Ford. Ford had 45 percent support, Ronald Reagan had 19, and other possible candidates followed. Most significantly, Gallup reported, a year before the convention was to be held, that only 3 percent had "no preference." At the same time, however, Pat Caddell conducted a poll in New Hampshire for a private client. Caddell reported 36 percent for Ford, 30 for Reagan, and 34 percent undecided. Gallup had masked the true size of the undecided bloc by offering people a "laundry list" of ten or more candidates and forcing them to state a choice.

Caddell's figures on the Democratic side were even more striking. Senator Henry Jackson had 14 percent and Sargent Shriver 10; no other candidate was in double figures. The pollster reported that 63 percent of New Hampshire Democrats were undecided. Similarly, less than two months before the 1976 Massachusetts primary, Caddell's figures showed that 41 percent of that state's Democrats and independents were still undecided, a percentage more than twice that received by any candidate! Simultaneously, both Gallup's and Harris' national polls were showing the undecided rate at the 5 percent level.

Caddell at the time questioned the common polling practice of forcing people who declare that they are undecided to say which way they are leaning. "Things simply haven't shaken down yet. A lot of people are reluctant to pick a candidate, and if you want an accurate reading of public opinion, you shouldn't push them right now." It is unfortunate that both Gallup and Harris lump the leaners together with the solid supporters, as that makes it impossible to know how deep a given candidate's strength runs. Caddell's polls actually gave a much more reliable picture of public opinion in the months before the 1976 primaries, but newspapers give such reports little attention because they are not dramatic.

Even when an election is imminent, the size and character of the undecided vote is crucial. If two weeks before a presidential election 6 or 8 percent of the voters are saying that they are still undecided, it would be most useful to know what portion of them are Democrats, independents, and Republicans; how they voted in prior elections; and where they stand on the principal issues of the campaign. Pollsters keep pushing to see which way people are leaning when it would be much more fruitful to discover why they have not been able to make up their minds.

Even without this information, the careful reader can learn something from the percentage of undecided voters. For example, if one candidate is better known than his opponent, then a relatively high percentage of undecideds may indicate that the better-known candidate has some problems. People already have formed an opinion about him; he has probably garnered most of the support he can get. The lesser-known candidate usually has a better chance to win the undecided vote, though of course it is not guaranteed to him.

The people who say that they have not made up their minds are not always telling the truth. If there is a controversial candidate—a George Wallace or a Barry Goldwater—the undecided percentage in part probably represents "hidden voters" who are afraid or unwilling to express their true preferences. This was particularly true in the 1972 primaries, in which Wallace almost invariably got a much higher share of the vote than the pre-election polls had shown. In Indiana, for example, he got 41 percent of the vote, not the 22 percent he had received in the polls. In Florida, he got 42 percent, not 34; in Wisconsin, 22 percent instead of 15; and in Massachusetts, 9 percent, not the 4 percent reported by the polls. The only exception to this pattern came in California, and there his name was not included on the ballot.

The pollsters are aware of this hidden-vote phenomenon, in Wallace's case in particular, but they have not been able to find a way to get people to be truthful with them.

5. Issue polls: could you answer the question? Polls on issues, be they foreign policy or domestic matters, present an entirely different problem for the pollster. For any given issue, there is a whole range of possible opinions, not just two. The more complex the issue is, the greater the range. Nevertheless, most pollsters try to fit all opinion into the neat categories of agree/disagree, favor-/oppose. These simplistic categories make it easy for the key-punch operators to code the results, and they also make for powerful headlines, but they mask the color and depth of public opinion as it truly exists.

Daniel Yankelovich is one of the few pollsters to acknowledge the problem. "The fact is, we have overgeneralized the pre-election model. The techniques which are correct for a pre-election survey aren't necessarily right for assessing a more complex kind of issue, one that people have not thought through." According to Yankelovich, the pollsters have yet to solve the problem of polling on issues. "I don't think the polling profession can pause to congratulate itself for what it has accomplished, nor can it present itself as being scientific, as long as this problem, which is immense, persists."

The best way to test a question to see whether it is a good indicator of public opinion is to see whether you would feel completely comfortable answering it yourself. If you find yourself saying "yes, but" or "no, except," then you must disregard the poll. The pollsters did not allow the fifteen hundred people they talked with to express their qualifications or reservations. Only part of their opinions are reflected in the poll.

The Gallup Poll makes it a practice to include the specific wording of all questions so that readers can make their own judgments about what the responses mean. Gallup has been criticized by a number of his colleagues for releasing his polls without making an analysis. It may be that the results can be confusing for the unsophisticated reader, but anyone who examines the wording carefully can interpret the results himself.

Harris, by contrast, does not include the wording of his questions—he says it would take too much space—so his readers have to take it on faith that he is interpreting the results correctly. When one reads the Harris Survey, one is reading not a poll but rather the political analysis of one individual.

In the best of all worlds, newspaper polls would include both the exact wording of the questions and an analysis as well. Readers would thus be warned that the results do not necessarily mean what they seem to say, and they still would have the information necessary to criticize the pollster's personal interpretation.

The newspaper polls on issues suffer from another important defect. Very rarely do they reveal whether people who know nothing about the issue were filtered out of the sample. It makes little sense to ask people how they feel about the president's policy on deregulation of natural gas prices if they have not previously considered the matter; a surprisingly high percentage of the population does not think about such questions. Indeed, the pollsters ought to be reporting less on whether the people support or oppose a given policy and more on whether they have even thought of it. There should also be much more done in the direction of showing intensities of opinion, measuring how deeply people care about a given issue.

Unfortunately, the newspaper polls do not give the reader much information on which to base an intelligent judgment about their reliability. Even if the wording of a particular question is given, and appears to be neutral, it is still possible that the poll was biased. One of the more subtle ways of rigging a poll is to arrange the questions in such a way that respondents will be inclined to give a certain answer. Only by seeing the entire survey can one really be sure of the meaning of answers to particular questions.

6. *What do the other polls say?* A properly done survey should include many different questions to get at all the nuances and facets of public opinion. All too often the national pollsters rely on only a few questions which give just the barest outlines of opinion. By looking at several polls together, the careful reader can sometimes get a truer picture of public attitudes. Robert Teeter, who has polled for both Nixon and Ford, warns: "Be very suspicious of any poll that claims to reveal new or startling results. If it's out of line with the other polls, it could very well be wrong."

A comparison of several polls can also provide a much-needed reminder of how elusive opinion is and how clumsy the pollsters often are in trying to trap it. In December 1975, for example, the Harris Survey reported that Hubert Humphrey held a 52 to 41 percent lead over Gerald Ford, but only three days later Gallup released a poll which had Ford trouncing Humphrey 51 to 39! The pollsters were at a loss to explain the contradictory figures. "It amazes me," Harris said. A Gallup spokesman stated, "There is no way that it can be accounted for."

A comparison between the polls can also show how rapidly opinion can change in response to changing events. In early 1967 the *Washington Post* printed, side by side, two seemingly contradictory polls about Robert Kennedy. The Gallup Poll reported that voting-age adults favored Kennedy over Lyndon Johnson for the presidency by 48 to 39 percent, with the remainder undecided. In the adjacent column, however, the Harris Survey said that among those who expressed an opinion, Johnson was preferred over Kennedy by a 56 to 44 ratio. George Gallup wrote, "Senator Robert Kennedy's star looms larger than ever on the political horizon," while Lou Harris concluded, "Senator Kennedy's standing with the public has just taken a tumble."

Some variation in the wording of the questions could have accounted for part of the difference, as could sampling error, but the most likely explanation lay elsewhere. Gallup's figures were based on interviews conducted two weeks before Harris', even though both polls were released to the newspapers on the same day. In the short time between the taking of the two polls, a highly publicized controversy arose between the Kennedy family on one side and author William Manchester and *Look* on the other. The Kennedys demanded that Manchester delete some material from his manuscript, *The Death of a President*, while he and his publishers insisted that it remain. Robert Kennedy was cast as the heavy in the dispute, and apparently it cost him some support in the Harris results.

The incident was soon forgotten and doubtless had no lasting effect on the public's impression of Robert Kennedy. Opinion is transitory, ever changing, like a cloud. In response to the slightest breeze, it can billow or evaporate. No matter what, it will be a different shape tomorrow.

7. What did the pollster really ask? The results of a survey vary significantly with rather inconspicuous changes in wording and format. The December 1975 Gallup Poll that catapulted Reagan into a 40 to 32 percent lead over Ford had presented people with a "laundry list" of ten Republicans, including many who had expressly declared their non-candidacy. The poll had a dramatic effect on press coverage of the campaign; Reagan was transformed from a rather quixotic challenger to the new frontrunner.

In part, however, Reagan's apparent lead may have simply been due to the way that Gallup had posed the question. Just three weeks later, in a far less publicized poll, Gallup asked people to choose between Reagan and Ford. This time the two candidates tied at 45 percent each, with 10 percent undecided. The second study may have been a more realistic look at the then forthcoming primaries, as it pitted the two principal contenders head to head. By contrast, any laundry-list question tends to reduce the percentage of undecided voters; it may also hurt centrist candidates, whose support is chipped away from both sides.

Similarly, as a Harris Survey released in early 1976 demonstrated, it is easy to read too much into the responses to any single question. When Harris presented Democrats and independents with a list of twenty-five possible candidates, Wallace was supported by 17 percent of the interviewees, second only to Kennedy; with Kennedy's name removed, Wallace got 19 percent, second to Humphrey. Jimmy Carter was far down on the list with a lowly 2 percent. It would be erroneous, however, to conclude that Wallace was strongly preferred over Carter. In fact, when Harris matched the two men against each other in a separate question, Carter topped Wallace 43 to 38 percent, with the rest undecided. Read together, the two sets of figures indicated that although Wallace's appeal was intense, it was quite limited.

8. Who paid for the poll? There have been flagrant examples of people sponsoring biased polls. Democratic financer Arthur Krim told Archibald Crossley where to poll and what questions to ask in order to come up with a poll which would make Lyndon Johnson look good. When publisher Ralph Ginzburg wanted to smear Barry Goldwater, he came up with his own loaded questions and biased samples.

Many pollsters will come through with a poll which says whatever the client wants, as long, of course, as the client is willing to pay for it. Charles W. Roll, Jr., and Albert H. Cantril, in their book *Polls: Their Use and Misuse in Politics*, tell of two incidents which reveal just how far some pollsters are willing to go. In one case, a pollster who was competing to be hired for a statewide race offered to present two surveys, one an accurate report of opinion, the other a phony poll to be released to the press. In the other case, a pollster actually tried to blackmail a prospective client, saying that unless he were hired, he would make him look bad by releasing a damaging survey to the press. The fact that a pollster or firm is well known does not mean that he does not have a price.

Reports by politicians on what their constituents tell them are completely unreliable. In May 1972, after Richard Nixon ordered the mining of North Vietnamese ports, administration spokesmen reported that public response was in support of the action by a five-to-one margin. Many people had indeed telephoned the White House to express their opinions, but the calls had been tabulated so as to inflate Nixon's support. When people told the switchboard operator that they wanted to register their views, she would ask them if they were in favor or opposed. If they were in favor, they would immediately be put down in the support column, but if they said they were opposed, they were put on hold while supposedly someone was found to record their opinion. After hanging on the line twenty minutes or so, most people just gave up.

Nixon certainly did not originate this gambit. Bill Moyers, former press secretary for Lyndon Johnson, recalls that one evening, after an important presidential speech on Vietnam, Moyers' wife telephoned to ask what time he would be home for dinner. She was recorded as having called to express support of Johnson's announced policies.

In a sense, there are no truly independent polls. Someone always pays for them and has an interest in having them made public. Even if newspapers give equal attention to polls which contradict their editorial policies (and many do not), they have a bias toward polls which depict the public as sharply divided; surveys which show that most people are undecided or do not care about an issue are usually ignored. Beware especially of leaked polls which find their way into political columns. These leaks are never accompanied by enough information to test their validity, and you can be sure that only the information which is favorable to the candidate is told to the press.

9. Is it really a trend? In late 1975 the *New York Times* headlined, "California Poll Has Ford Losing Ground to Reagan." An examination of the survey, however, shows that conclusion to be virtually groundless. In August the Republican contenders had been in a dead heat, while in November Reagan was reported to have a single-point lead. The sampling error for the survey, however, was plus or minus 5.5 points for each candidate, so it was within the realm of normal chance that Ford actually had a ten-point lead! The newspapers painstakingly keep track of each little dip or rise in a candidate's standings, but such shifts have no real statistical significance.

Charting of trends is the common mode of analysis in newspaper polls; this is particularly true of the Gallup Poll. Even where the changes themselves may be statistically significant, this emphasis on tracing the peaks and valleys over time may be extremely misleading. In our fascination with the graph we lose sight of what is really being measured.

In 1975, successive Gallup polls had approval of President Ford's performance in office skittering all over the chart. In January it was an anemic 37 percent. In early March, it appeared to be up to 39, but at the end of the month, it was back to 37. In the beginning of April it shot up seven points to 44, but within weeks it went down to 39. It was 40 in May, then rocketed up to 51 in early June and 52 at the end of that month. A month later it was back down to 45.

Can we really believe that these numbers mean anything? The pollsters would have us think that they do. Ford's rating shot up after he retaliated with military force for the Cambodian seizure of the *Mayaguez*. But his apparent gains proved to be merely temporary. The Gallup approval ratings reflect superficial and volatile moods, but they tell us next to nothing about bedrock support of a president's overall performance. The drama of watching the president's stock rise or plummet, however, is exciting, so we end up looking at just one indicator of support when we should be looking at a whole host of them.

The pollsters have to push this trend analysis because it provides them with an ever-present alibi for their polls. If there is a difference between two successive polls, it is not that one of them could possibly be wrong but rather that "each was an accurate picture of opinion at the time it was taken." It also provides a never-ending market for their services, for no poll can ever be the last word.

Gallup and Harris have been accused of secretly arranging to release their polls, particularly those on elections, on different days so that neither one will be embarrassed if the results differ. Even if the surveys were taken only a few days apart, any differences can be explained away by the argument that public opinion changed over that period.

Both Gallup and Harris emphatically deny the accusation,

and, given their ill-concealed antipathy toward each other, it is hard to believe that they could agree on anything. Even if they do not conspire, the fact is that their polls rarely do appear simultaneously. If that is a coincidence, it is one which both men must be happy to live with. If they really wanted to stifle the accusations of self-serving conspiracy, they would simply announce that for one election they would arrange to poll at the same time. Such an experiment would put both pollsters to the test, and would provide an interesting demonstration of just how scientific polling actually is.

10. How much do you know about the poll? If you keep the foregoing questions in mind when you read a poll, you may be left with the uneasy feeling that you do not know enough about it to make an intelligent judgment as to whether it is reliable. That is good. The less you know about the poll—the way it was conducted, the wording of the questions, who sponsored it, and so on—the less you should trust it.

Newspapers have recently been more frequently outlining certain basic technical information about the polls they publish, but we are seldom told how the pollster decided which of the people he talked with are likely to vote and which are not. Nor are we told whether filter questions were used. When we get beyond straight technical questions, we must always operate in the dark. Gallup provides the wording of the questions he refers to, but we do not see the order in which the questions were asked.

When opinions are squeezed to fit the polling process, they lose their shape and vitality. Figures in the newspapers seem so precise and unequivocal that they belie the fact that public opinion is a multifaceted and changing mass. The fundamental flaw of most polls is that they describe in black and white something which is so manyhued and so complex that it never can be reduced to just two tones, in only two dimensions.

Readers may find the cautious tone of this advice unsatisfactory. It would be more exciting to know the ten tips on how to decode public opinion polls and dope out an election. We want to understand public opinion and, even more, we want to believe that the process of choice and policy-making is as rational, measurable, and certain as the polls make them seem. We have willingly let the polls remake politics in their own image. The unhappy truth, however, is that if we look to the polls uncritically for easy answers, we will end up misled more often than not.

A properly constructed poll can be a powerful tool for understanding public opinion. A poor one only makes matters worse. For every good poll there are dozens, perhaps hundreds, of bad ones. Unless you know enough about the poll to tell whether it is good or bad, the odds are that you are far better off being skeptical. Furthermore, any poll, good or bad, carries subtle implications about the way people think and the way decision-makers should respond which we must not accept uncritically.

Suggested Reading

Some people find the subject of public opinion polling, particularly its technical aspects, rather forbidding. To make this book accessible to such readers, I have omitted footnotes, scholarly references, and a formal bibliography. For those who have overcome their trepidation and wish to pursue the topic, I have listed a number of books which deal with various phases of public opinion polling and provide an entry into the rather considerable literature on the subject.

Charles H. Backstrom and Gerald D. Hursh have written a useful primer, *Survey Research* (Evanston: Northwestern University Press, 1963), for those who wish either to conduct their own poll or simply to know more about survey techniques. Though brief, Darrell Huff's *How to Lie with Statistics* (New York: W. W. Norton, 1954) is a most effective—and entertaining—explanation of how we are often bamboozled by numbers.

In recent years there have been quite a few books on professional campaign consultants. *The Election Men*, by David Lee Rosenbloom (New York: Quadrangle Books, 1973) describes, among other things, the use of polls in modern politicking. A number of examples of polling chicanery are documented in *Polls: Their Use and Misuse in Politics*, by Charles W. Roll, Jr., and Albert H. Cantril (New York: Basic Books, 1972).

Though published almost three decades ago, Lindsay Rogers' *The Pollsters* (New York: Alfred A. Knopf, 1949) remains the most erudite examination of public opinion polls and their proper role in a democracy. *Silent Politics Polls and the Awareness of Public Opin-*

ion, by Leo Bogart (New York: Wiley-Interscience, 1972) fleshes out Rogers' thesis with examples drawn from contemporary polls. Robert Nesbit's "Public Opinion versus Popular Opinion" (*The Public Interest*, 41, Fall 1975, p. 166) is also most enlightening.

There has been relatively little published about George Gallup and Louis Harris, but much can be gleaned from what they have written about polling. Though Gallup has produced several books more recently, his *Pulse of Democracy,* co-authored with Saul Rae (New York: Simon & Schuster, 1940) reflects most clearly his fundamental views. Harris' *The Anguish of Change* (New York: W. W. Norton, 1973), presents his view of American opinion and his role in reporting it. Both Gallup and Harris say that their firms will answer any inquiry about polls mentioned in their syndicated newspaper columns.

The *Public Opinion Quarterly*, published by the American Association for Public Opinion Research, includes noteworthy surveys and articles on polling techniques. The Roper Public Opinion Research Center at Williams College, Williamstown, Massachusetts, is probably the best repository of opinion polling information.

Index

96   Index